Praise for *The Intention Experiment*

'If you want to explore the latest science behind *The Secret*, look no further. Science and wisdom collide and make friends in this real-world adventure that is ultimately a guidebook for living.'
Drew Heriot, director of The Secret

'Lynne McTaggart has zeroed in on a wonderful collection of experiments and events that shatters our normal materialistic assumptions of time, space, and everything in between (if there *is* an in-between). It's as mind-bending as it's meant to be.'
William Arntz, producer, writer, and director of What the BLEEP Do We Know!? *and author of* What the Bleep do We Know!? Discovering the Endless Possibilities for Altering Your Everyday Reality.

'Quantum physicists, myself included, are predicting for some time that our intentions are powerful when made in the context of nonlocal consciousness. In this book Lynne McTaggart demonstrates this with empirical data. It will convince you of the power of intention and the power of nonlocal consciousness. A very important book for the aborning science within consciousness.'
Amit Goswami, Professor of physics, emeritus, University of Oregon, and author of The Self Aware Universe *and* The Quantum Doctor

'Lynne McTaggart has a dazzling genius for bringing together cutting-edge research in the field of quantum physics in a stunningly direct, accessible way. *The Intention Experiment* is not just a fascinating read, it's an exciting invitation to begin your own experiment to experience directly the profound influence intention has in your life.'
Brandon Bays, author of The Journey *and* Freedom Is

For Anya
A master of intention

About the Author

LYNNE McTAGGART IS AN award-winning journalist and author of the bestselling book *The Field* (www.livingthefield.com). As co-founder and editorial director of What Doctors Don't Tell You (www.wddty.co.uk), she publishes health newsletters that are among the most widely praised in the world. She is also editor of *Living The Field*, a course that helps to bring the science of *The Field* into every-day life. Her company also holds highly popular conferences and workshops on health and spirituality.

She has become a well-respected international authority on the science of spirituality.

Ms McTaggart is also author of *The Baby Brokers: The Marketing of White Babies in America* (The Dial Press) and *Kathleen Kennedy: Her Life and Times* (The Dial Press/Weidenfeld & Nicolson in the UK). *The Field* and *What Doctors Don't Tell You* each have been translated into many languages around the world.

She and her husband, WDDTY co-founder Bryan Hubbard, live and work in London with their two daughters.

To visit Lynne at her website, see www.theintentionexperiment.com.

For Living the Field Master Class, conferences, workshops and products, see www.livingthefield.com. For her health publications, see www.wddty.com.

USE YOUR THOUGHTS
TO CHANGE THE WORLD

The Intention
Experiment

LYNNE McTAGGART

HarperElement
An Imprint of HarperCollins*Publishers*
77–85 Fulham Palace Road,
Hammersmith, London W6 8JB

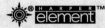

and *HarperElement* are trademarks of
HarperCollins*Publishers* Ltd

First published by HarperElement 2007
This paperback edition published 2008

7

© Lynne McTaggart 2007, 2008

Lynne McTaggart asserts the moral right to
be identified as the author of this work

A catalogue record of this book is
available from the British Library

ISBN 978-0-00-719459-9

Mixed Sources
Product group from well-managed
forests and other controlled sources
www.fsc.org Cert no. SW-COC-1806
© 1996 Forest Stewardship Council
FSC

Contents

God is afoot, magic is alive
… magic never died

Magic is Alive, Leonard Cohen

Acknowledgements

THE INTENTION EXPERIMENT has been assembled from multiple interviews or correspondence with most of the scientists or medical doctors described in its pages, plus a close reading of their major scientific papers. These include: Harald Atmanspacher, Cleve Backster, Dick Bierman, Caslav Brukner, Melinda Connor, Eric Davis, Richard Davidson, John Diamond, Walter Dibble, Thomas Durt, Sayantani Ghosh, Stuart Hameroff, Valerie Hunt, Mitch Krucoff, Konstantin Korotkov, Stanley Krippner, Sarah Lazar, Leonard Leibovici, Todd Murphy, Roger Nelson, Michael Persinger, Fritz-Albert Popp, Dean Radin, Benni Reznik, Thomas Rosenbaum, Metod Saniga, Marilyn Schlitz, Gary Schwartz, Jerome Stone, Ingo Swann, William Tiller, Eduard Van Wijk and Fred Alan Wolf.

I also interviewed a number of people trained or gifted in the art of intention: clairvoyants such as Ingo Swann, *Qigong* experts such as Bruce Kumar Frantzis, healers such as Eric Pearl, plus a number of other experienced healers who filled out an extensive questionnaire.

I am especially grateful to Vlatko Vedral, who schooled me in the latest quantum theory, Gary Schwartz, for his many inventive ideas and help on many fronts, William Tiller, who has explained his theories painstakingly, Stanley Krippner, who has provided many leads and much research, and Dean Radin, for his extra help on the science of retro-intention. I owe a great debt to Cleve Backster, Dick Bierman, Caslav Brukner, Richard Davidson, Sai Ghosh, Konstantin Korotkov, Stanley Krippner, Todd Murphy, Michael Persinger, Fritz-Albert Popp, Dean Radin, Thomas Rosenbaum, Gary Schwartz, Jerome Stone, William Tiller, Eduard Van Wijk and Vlatko Vedral, all of whom read the portions of the manuscript that describe their work and corrected any errors. Among books, I am indebted to Larry Dossey's *Be Careful What You Pray For* and *Healing Words*, Marilyn

Schlitz's excellent compilation *Consciousness and Healing*, Daniel Benor's several books and excellent website, the books of William Tiller, Dean Radin's *Entangled Minds*, and *Primary Perception* by Cleve Backster. Various bibliographies published on the Internet were particularly helpful, including: Radin's bibliography from *Entangled Minds*, Michael Murphy's bibliography from *The Science of Meditation*, and the bibliographies of Stephan Schwartz and the Retro-PK project.

Particular thanks are due to Suzanne Donahue, Heidi Metcalfe, Shannon Gallagher and Andrew Paulson at Free Press, and Wanda Whiteley, Liz Dawson and Belinda Budge at HarperCollins in the UK for raising the bar on this project and supporting it at every stage. My editors Leslie Meredith and Katy Carrington, and my copyeditors, the eagle-eyed Andrew Coleman in the UK and 'Violent' Viola and Bryan Cholfin in America, improved this manuscript in countless ways.

A special mention should be made of Will Arntz, Betsy Chasse and Mark Vicente, producers of *What the Bleep Do We Know!?*, and their ongoing support of *The Field* and my other projects. And of my entire team at my company Conatus, particularly Tony Edwards, Joanna Evans, Nicolette Vuvan and Pavel Mikoloski, who are most involved with *Living The Field*.

My agents Russell Galen and Daniel Baror once more demonstrated an extraordinary commitment to this project from the start and worked tirelessly to find it a good home around the world.

I am grateful for what I learn every day from my children, Caitlin and Anya, about the extraordinary power of intention.

The contributions of Robert Jahn, Brenda Dunne and Fritz-Albert Popp, Eduard Van Wijk, Sophie Cohen, 'Annemarie' and all the staff at the International Institute for Biophysics in Germany who set up the first intention experiment, has been incalculable. This book truly would not exist without them.

And finally, this book owes its greatest debt, as always, to my husband, Bryan Hubbard, both for planting the first seed and then carefully nurturing it as it grew.

Preface

THIS BOOK REPRESENTS A PIECE of unfinished business that began in 2001 when I published a book called *The Field*. In the course of trying to find a scientific explanation for homeopathy and spiritual healing, I had inadvertently uncovered the makings of a new science.

During my research, I stumbled across a band of frontier scientists who had spent many years re-examining quantum physics and its extraordinary implications. Some had resurrected certain equations regarded as superfluous in standard quantum physics. These equations, which stood for the Zero Point Field, concerned the extraordinary quantum field generated by the endless passing back and forth of energy between all subatomic particles. The existence of the Field implies that all matter in the universe is connected on the subatomic level through a constant dance of quantum energy exchange.

Other evidence demonstrated that, on the most basic level, each one of us is also a packet of pulsating energy constantly interacting with this vast energy sea.

But the most heretical evidence of all concerned the role of consciousness. The well-designed experiments conducted by these scientists suggested that consciousness is a substance outside the confines of our bodies – a highly ordered energy with the capacity to change physical matter. Directing thoughts at a target seemed capable of altering machines, cells and, indeed, entire multicelled organisms like human beings. This mind–over–matter power even seemed to traverse time and space.

In *The Field* I aimed to make sense of all the ideas resulting from these disparate experiments and to synthesize them into one generalized theory. *The Field* created a picture of an interconnected universe and a scientific explanation for many of the most profound human mysteries, from alternative medicine and spiritual healing to extrasensory perception and the collective unconscious.

The Field apparently hit a nerve. I received hundreds of letters from readers who told me that the book had changed their lives. A writer wanted to depict me as a character in her novel. Two composers wrote musical compositions inspired by it, one of which was played on the international stage. I was featured in a movie, *What the Bleep!? Down the Rabbit Hole*, and on the *What The Bleep Do We Know!? Calendar*, released by the film's producers. Quotes from *The Field* became the centrepiece of a printed Christmas card.

However gratifying this reaction, I felt that my own journey of discovery had hardly left the station platform. The scientific evidence I had amassed for *The Field* suggested something extraordinary and even disturbing: directed thought had some sort of central participatory role in creating reality.

Targeting your thoughts – or what scientists ponderously refer to as 'intention' and 'intentionality' – appeared to produce an energy potent enough to change physical reality. *A simple thought seemed to have the power to change our world.*

After writing *The Field*, I puzzled over the extent of this power and the numerous questions it raised. How, for instance, could I translate what had been confirmed in the laboratory for use in the world that I lived in? Could I stand in the middle of a railway track and, Superman-style, stop the 9:45 to Paddington with my thoughts? Could I fly myself up to fix my roof with a bit of directed thought? Would it now be possible to cross doctors and healers off my list of essential contacts, seeing as I might now be able to think myself well? Could I help my children pass their maths tests just by thinking about it? If linear time and three-dimensional space didn't really exist, could I go back and erase all those moments in my life that had left me with lasting regret? And could my one puny bit of mental input do anything to change the vast catalogue of suffering on the planet?

The implications of this evidence were unsettling. Should we be minding every last thought at every moment? Was a pessimist's view of the world likely to be a self-fulfilling prophecy? Were all those negative thoughts – that ongoing inner dialogue of judgement and criticism – having any effect outside our heads?

Were there conditions that improved your chances of having a better effect with your thoughts? Would a thought work any old time or would you, your intended target and indeed the universe itself have to be in the mood? If everything is affecting everything else at every moment, doesn't that counteract and thereby nullify any real effect?

What happens when a number of people think the same thought at the same time? Would that have an even larger effect than thoughts generated singly? Was there a threshold size that a group of like-minded intenders had to reach in order to exert the most powerful effect? Was an intention 'dose dependent' – the larger the group, the larger the effect?

An enormous body of literature, starting with *Think and Grow Rich*,[1] by Napoleon Hill, arguably the first self-actualization guru, has been generated about the power of thought. 'Intention' has become the latest New Age buzzword. Practitioners of alternative medicine speak of helping patients heal 'with intention'. Even Jane Fonda writes about raising children 'with intention'.[2]

What on earth, I wondered, was meant by 'intention'? And how exactly can one become an efficient 'intender'? The bulk of the popular material had been written off the cuff – a smattering of Eastern philosophy here, a *soupçon* of Dale Carnegie there – with very little scientific evidence that it worked.

To find answers to all of these questions, I turned, once again, to science, scouring the scientific literature for studies on distant healing or other forms of psychokinesis, or mind over matter. I sought out international scientists who experimented with how thoughts can affect matter. The science described in *The Field* had been carried out mainly in the 1970s; I examined more recent discoveries in quantum physics for further clues.

I also turned to those people who had managed to master intention and who could perform the extraordinary – spiritual healers, Buddhist monks, *Qigong* masters, shamans – in order to understand the transformational processes they underwent to be able to use their thoughts to powerful effect. I uncovered myriad ways that intention is used in real life – in sports, for instance, and during healing modalities such as biofeedback. I studied how native populations incorporated directed thought into their daily ritual.

I then began to dig up evidence that multiple minds trained on the same target magnified the effect produced by an individual. The evidence was tantalizing, mostly gathered by the Transcendental Meditation organization, suggesting that a group of likeminded thoughts created some sort of order in the otherwise random Zero Point Field.

At that point in my journey, I ran out of pavement. All that stretched before me, as far as I could tell, was uninhabited open terrain.

Then one evening, my husband Bryan, a natural entrepreneur in most situations, put forward what seemed to be a preposterous suggestion: 'Why don't you do some group experiments yourself?'

I am not a physicist. I am not any kind of scientist. The last experiment I had conducted had been in a 10th grade science lab.

What I did have, though, was a resource available to few scientists: a potentially huge experimental body. Group intention experiments are extraordinarily difficult to perform in an ordinary laboratory. A researcher would need to recruit thousands of participants. How would he find them? Where would he put them? How would he get them all to think the same thing at the same time?

A book's readers offer an ideal self-selected group of likeminded souls who might be willing to participate in testing out an idea. Indeed, I already had my own large population of regular readers with whom I communicated through e-news and my other spin-off activities from *The Field*.

I first broached the idea of carrying out my own experiment with dean emeritus of the Princeton University School of Engineering Robert Jahn and his colleague, psychologist Brenda Dunne, who run the Princeton Engineering Anomalous Research (PEAR) laboratory, both of whom I had got to know through my research for *The Field*. Jahn and Dunne have spent some 30 years painstakingly amassing some of the most convincing evidence about the power of directed intention to affect machinery. They are absolute sticklers for scientific method, no-nonsense and to the point. Robert Jahn is one of the few people I have ever met who speaks in perfect, complete sentences. Brenda Dunne is equally perfectionist about detail in both experiment and language. I would be assured of no sloppy protocol in my experiments if Jahn and Dunne agreed to be involved.

The two of them also have a vast array of scientists at their disposal. They head the International Consciousness Research Laboratory, many of whose members are among the most prestigious scientists performing consciousness research in the world. Dunne also runs PEARTree, a group of young scientists interested in consciousness research.

Jahn and Dunne immediately warmed to the idea. We met on numerous occasions and kicked around some possibilities. Eventually, they put forward Fritz-Albert Popp, assistant director of the International Institute of Biophysics (IIB) in Neuss, Germany, to conduct the first intention experiments. I knew Fritz Popp through my research for *The Field*. He was the first to discover that all living things emit a tiny current of light. As a noted German physicist recognized internationally for his discoveries, Popp would also be a stickler for pristine scientific method.

Other scientists, such as psychologist Gary Schwartz of the Biofield Center at the University of Arizona, Marilyn Schlitz, vice president for research and education at the Institute of Noetic Sciences, Dean Radin, IONS' senior scientist, and psychologist Roger Nelson of the Global Consciousness Project, have also offered to participate.

I do not have any hidden sponsors of this project. The website and all our experiments will be funded by the proceeds of this book or grants, now and in the future.

Scientists involved in experimental research often cannot venture beyond their findings to consider the implications of what they have uncovered. Consequently, when assembling the evidence that already exists about intention, I have tried to consider the larger implications of this work and to synthesize these individual discoveries into a coherent theory. In order to describe in words concepts that are generally depicted through mathematical equations, I have had to reach for metaphoric approximations of the truth. At times, with the help of many of the scientists involved, I have also had to engage in speculation. It is important to recognize that the conclusions arrived at in this book represent the fruits of frontier science. These ideas are a work in progress. Undoubtedly new evidence will emerge to amplify and refine these initial conclusions.

Researching the work of people at the very forefront of scientific discovery again has been a humbling experience for me. Within the unremarkable confines of a laboratory, these largely unsung men and women engage in activities that are nothing short of heroic. They risk losing grants, academic posts and, indeed, entire careers groping alone in the dark. Most scratch around for grant money to enable them to carry on.

All advancements in science are somewhat heretical, each important new discovery partly, if not completely, negating the prevailing views of the day. To be a true explorer in science – to follow the unprejudiced lead of pure scientific inquiry – is to be unafraid to propose the unthinkable, and to prove friends, colleagues and scientific paradigms wrong. Hidden within the cautious, neutral language of experimental data and mathematical equation is nothing less than the makings of a new world, which slowly takes shape for all the rest of us, one painstaking experiment at a time.

Lynne McTaggart, June 2006

Introduction

THE INTENTION EXPERIMENT is no ordinary book, and you are no ordinary reader. This is a book without an ending, for I intend for you to help me finish it. You are not only the audience of this book, but also one of its protagonists – the primary participants in cutting-edge scientific research. You, quite simply, are about to embark on the largest mind–over–matter experiment in history.

The Intention Experiment is the first 'living' book in three-dimensions. The book, in a sense, is a prelude, and the 'contents' carry on well beyond the time you finish the final page. In the book itself, you will discover scientific evidence about the power of your own thoughts, and you will then be able to extend beyond this information and test further possibilities through a massive, ongoing international group experiment, under the direction of some of the most well-respected international scientists in consciousness research. Through *The Intention Experiment*'s website (www.theintention experiment.com), you and the rest of the readers of this book will be able to participate in remote experiments, the results of which will be posted on the site. Each of you will become a scientist at the hub of some of the most daring consciousness experiments ever conducted.

The Intention Experiment rests on an outlandish premise: thought affects physical reality. A sizeable body of research exploring the nature of consciousness, carried on for more than 30 years in prestigious scientific institutions around the world, shows that thoughts are capable of affecting everything from the simplest machines to the most complex living beings.[1] This evidence suggests that human thoughts and intentions are an actual physical 'something' with the astonishing power to change our world. Every thought we have is a tangible energy with the power to transform. A thought is not only a thing; a thought is a thing that influences other things.

This central idea, that consciousness affects matter, lies at the very heart of an irreconcilable difference between the world view offered by classical physics – the science of the big, visible world – and that of quantum physics – the science of the world's most diminutive components. That difference concerns the very nature of matter and the ways it can be influenced to change.

All of classical physics, and indeed the rest of science, is derived from the laws of motion and gravity developed by Isaac Newton in his *Principia*, published in 1687.[2] Newton's laws described a universe in which all objects moved within the three-dimensional space of geometry and time according to certain fixed laws of motion. Matter was considered inviolate and self-contained, with its own fixed boundaries. Influence of any sort required something physical to be done to something else – a force or collision. Making something change basically entailed heating it, burning it, freezing it, dropping it or giving it a good swift kick.

Newtonian laws, science's grand 'rules of the game', as the celebrated physicist Richard Feynman once referred to them,[3] and their central premise, that things exist independently of each other, underpin our own philosophical view of the world. We believe that all of life and its tumultuous activity carries on around us, regardless of what we do or think. We sleep easy in our beds at night, in the certainty that when we close our eyes, the universe doesn't disappear.

Nevertheless, that tidy view of the universe as a collection of isolated, well-behaved objects got dashed in the early part of the twentieth century, once the pioneers of quantum physics began peering closer into the heart of matter. The tiniest bits of the universe, those very things that make up the big, objective world, did not in any way behave themselves according to any rules that these scientists had ever known.

This outlaw behaviour was encapsulated in a collection of ideas that became known as the Copenhagen Interpretation, after the place where the forceful Danish physicist Niels Bohr and his brilliant protégé, the German physicist Werner Heisenberg, formulated the likely meaning of their extraordinary mathematical discoveries. Bohr and Heisenberg realized that atoms are not little solar systems of billiard balls but something far more messy: a tiny cloud of probability. Every subatomic particle is not a solid and stable thing, but exists simply as a potential of any one of its future selves – or what is known by physicists as a 'superposition', or sum, of all probabilities, like a person staring at himself in a hall of mirrors.

One of their conclusions concerned the notion of 'indeterminacy'; that you can never know all there is to know about a subatomic particle all at the same time. If you discover information about where it is, for instance, you cannot work out at the same time exactly where it is going or at what speed. They spoke about a quantum particle as both a particle – a congealed, set thing – and a 'wave function' – a big smeared-out region of space and time, any corner of which the particle may occupy. It was akin to describing a person as comprising the entire street where he lives.

Their conclusions suggested that, at its most elemental, physical matter isn't solid and stable – indeed, isn't an *anything* yet. Subatomic reality did not resemble the solid and reliable state of being described to us by classical science, but an ephemeral prospect of seemingly infinite options. So capricious seemed the smallest bits of nature that the first quantum physicists had to make do with a crude symbolic approximation of the truth – a mathematical range of all possibility.

At the quantum level, reality resembled unset jelly.

The quantum theories developed by Bohr, Heisenberg and a host of others rocked the very foundation of the Newtonian view of matter as something discrete and self-contained. They suggested that matter, at its most fundamental, could not be divided into independently existing units and indeed could not even be fully described. Things had no meaning in isolation, but only in a web of dynamic interrelationship.

The quantum pioneers also discovered the astonishing ability of quantum particles to influence each other, despite the absence of all those usual things that physicists understand are responsible for influence, such as an exchange of force occurring at a finite velocity. Once in contact, particles retained an eerie remote hold over each other. The actions – for instance, the magnetic orientation – of one subatomic particle instantaneously influenced the other, no matter how far they were separated.

At the subatomic level, change also resulted through dynamic shifts of energy; these little packets of vibrating energy constantly traded energy back and forth to each other like ongoing passes in a game of basketball, a ceaseless to-ing and fro-ing that gave rise to an unfathomably large basic layer of energy in the universe.[4]

Subatomic matter appeared to be involved in a continual exchange of information, causing constant refinement and subtle alteration. The universe was not a storehouse of static, separate objects, but a single organism of interconnected energy fields in a constant state of becoming. At its infinitesimal level, our world resembled a vast network of quantum information, with all its component parts constantly on the phone.

The only thing dissolving this little cloud of probability into something solid and measurable was the involvement of an observer. Once these scientists decided to have a closer look at a subatomic particle by taking a measurement, the subatomic entity that existed as pure potential would 'collapse' into one particular state.

The implications of these early experimental findings were profound: living consciousness somehow was the influence that turned the possibility of something into something real. The moment we looked at an electron or took a measurement, *it appeared that we helped to determine its final state*. This suggested that the most essential ingredient in creating our universe is the consciousness that observes it. Several of the central figures in quantum physics argued that the universe was democratic and participatory – a joint effort between observer and observed.[5]

The observer effect in quantum experimentation gives rise to another heretical notion: that living consciousness is somehow central to this process of transforming the unconstructed quantum world into something resembling everyday reality. It suggests not only that the observer brings the observed into being, but also that nothing in the universe exists as an actual 'thing' independently of our perception of it.

It implies that observation – the very involvement of consciousness – gets the jelly to set.

It implies that reality is not fixed, but fluid, or mutable, and hence possibly open to influence.

The idea that consciousness creates and possibly even affects the physical universe also challenges our current scientific view of consciousness, which developed from the theories of the seventeenth-century philosopher René Descartes – mind is separate and somehow different from matter – and eventually embraced the notion that consciousness is entirely generated by the brain and remains locked up in the skull.

Most modern workaday physicists shrug their shoulders over this central conundrum: that big things are separate, but the tiny building blocks they are made up of are in instant and ceaseless communication with each other. For half a century, physicists have accepted, as though it makes perfect sense, that an electron behaving one way subatomically somehow transmutes into 'classical' (that is, Newtonian) behaviour once it realizes it is part of a larger whole.

In the main, scientists have stopped caring about the troublesome questions posed by quantum physics, and left unanswered by its earliest pioneers. Quantum theory works mathematically. It offers a highly successful recipe for dealing with the subatomic world. It helped to build atomic bombs and lasers, and to deconstruct the nature of the sun's radiation. Today's physicists have forgotten about the observer effect. They content themselves with their elegant equations and await the formulation of unified Theory of Everything or the discovery of a few more dimensions beyond the ones that ordinary humans perceive, which they hope will somehow pull together all these contradictory findings into one centralized theory.

Thirty years ago, while the rest of the scientific community carried on by rote, a small band of frontier scientists at prestigious universities around the globe paused to consider the metaphysical implications of the Copenhagen Interpretation and the observer effect.[6] If matter was mutable, and consciousness *made* matter a set something, it seemed likely that consciousness might also be able to nudge things in a particular direction.

Their investigations boiled down to a simple question: if the act of *attention* affected physical matter, what was the effect of *intention* – of deliberately attempting to make a change? In our act of participation as an observer in the quantum world, we might be not only creators, but also influencers.[7]

They began designing and carrying out experiments, testing what they gave the unwieldy label of 'directed remote mental influence' or 'psychokinesis', or, in shorthand, 'intention' or even 'intentionality'. A textbook definition of intention characterizes it as 'a purposeful plan to perform an action, which will lead to a desired outcome',[8] unlike a desire, which means simply focusing on an outcome, without a purposeful plan of how to achieve it. An intention was directed at the intender's own actions; it required some sort of reasoning; it required a commitment to do the intended deed. Intention implied purposefulness: an understanding of a plan of action and a planned satisfactory result. Marilyn Schlitz, vice-president for research and education at the Institute of Noetic Sciences and one of the scientists engaged in the earliest investigations of remote influence, defined intention as 'the projection of awareness, with purpose and efficacy, toward some object or outcome'.[9] To influence physical matter, they believed, thought had to be highly motivated and targeted.

In a series of remarkable experiments, these scientists provided evidence that thinking certain directed thoughts could affect one's own body, inanimate objects and virtually all manner of living things, from single-celled organisms to human beings. Two of the major figures in this tiny subgroup were former dean of engineering Robert Jahn at the Princeton Anomalies Engineering Research (PEAR) laboratory at Princeton University and his colleague Brenda Dunne, who together created a sophisticated, scholarly research programme grounded in hard science. Over 25 years, Jahn and Dunne led what became a massive international effort to quantify what is referred to as 'micro-psychokinesis', the effect of mind on random-event generators (REGs), which perform the electronic, twenty-first century equivalent of a toss of a coin.

The output of these machines (the computerized equivalent of heads or tails) was controlled by a randomly alternating frequency of positive and negative pulses. Because their activity was utterly random, they produced 'heads' and 'tails' each roughly 50 per cent of the time, according to the laws of probability. The most common configuration of the REG experiments was a computer screen randomly alternating two attractive images – say, of cowboys and Indians. Participants in the studies would be placed in front of the computers and asked to try to influence the machine to produce more of one image – more cowboys, say – then to focus on producing more images of Indians, and then to try not to influence the machine in either direction.

Over the course of more than two and a half million trials Jahn and Dunne decisively demonstrated that human intention can influence these electronic devices in the specified direction,[10] and their results were replicated independently by 68 investigators.[11]

While PEAR concentrated on the effect of mind on inanimate objects and processes, many other scientists experimented with the effect of intention on living things. A diverse number of researchers demonstrated that human intention can affect an enormous variety of living systems: bacteria, yeast, algae, lice, chicks, mice, gerbils, rats, cats and dogs.[12] A number of these experiments have also been carried out with human targets; intention has been shown to affect many biological processes within the receiver, including gross motor movements and those in the heart, the eye, the brain and the respiratory system.

Animals themselves proved capable of acts of effective intention. In one ingenious study by René Peoc'h of the Fondation ODIER in Nantes, France, a robotic 'mother hen', constructed from a moveable random-event generator, was 'imprinted' on a group of baby chicks soon after birth. The robot was placed outside the chicks' cage, where it moved around freely, as its path was tracked and recorded. Eventually, it was clear that the robot was moving towards the chicks two and a half times more often than it would ordinarily; the 'inferred intention' of the chicks – their desire to be close to their mother – appeared to affect the robot, drawing it closer to the cage. In over 80 similar studies, in which a lighted candle was placed on a movable REG, baby chicks kept in the dark, finding the light comforting, managed to influence the robot to spend more time than normal in the vicinity of their cage.[13]

The largest and most persuasive body of research has been amassed by William Braud, a psychologist and the research director of the Mind Science Foundation in San Antonio, Texas, and, later, the Institute of Transpersonal Psychology. Braud and his colleagues demonstrated that human thoughts can affect the direction in which fish swim, the movement of other animals such as gerbils, and the breakdown of cells in the laboratory.[14]

Braud also designed some of the earliest well-controlled studies of mental influence on human beings. In one group of studies, Braud demonstrated that one person could affect the autonomic nervous system (or fight-or-flight mechanisms) of another.[15] Electrodermal

activity (EDA) is a measure of skin resistance and shows an individual's state of stress; a change of EDA usually occurs if someone is
stressed or made uncomfortable in some way.[16] Braud's signature
study tested the effect on EDA of being stared at, one of the simplest
means of isolating the effect of remote influence on a human being.
He repeatedly demonstrated that people were subconsciously
aroused while they were being stared at.[17]

Perhaps the most frequently studied area of remote influence
concerns remote healing. Some 150 studies, of variable scientific
rigour, have been carried out,[18] and one of the best designed was
conducted by the late Dr Elisabeth Targ. During the height of the
AIDS epidemic in the 1980s, she devised an ingenious, highly
controlled pair of studies, in which some 40 remote healers across
America were shown to improve the health of terminal AIDS
patients, even though the healers had never met or been in contact
with their patients.[19]

Even some of the most rudimentary mind–over–matter experiments have had tantalizing results. One of the first such studies
involved attempts to influence a throw of the dice. To date, some 73
studies have examined the efforts of 2500 people to influence more
than two and a half million throws of the dice, with extraordinary
success. When all the studies were analysed together, and allowances
made for quality or selective reporting, the odds of the results occurring by chance alone were 10^{76} (1 followed by 76 zeros) to one.[20]

There was also some provocative material about spoon bending,
that perennial party trick made popular by psychic Uri Geller. John
Hasted, a professor at Birkbeck College at the University of London,
had tested this with an ingenious experiment involving children.
Hasted suspended latch keys from the ceiling and placed the children
3 to 10 feet away from their target key, so that they could have no
physical contact. Attached to each key was a strain gauge, which
would detect and register on a strip chart recorder any change in the
key. Hasted then asked the children to try to bend the suspended

metal. During the sessions, he observed not only the keys swaying and sometimes fracturing, but also abrupt and enormous spikes of voltage pulses up to 10 volts – the very limits of the chart recorder. Even more compelling, when children had been asked to send their intention to several keys hung separately, the individual strain recorders noted simultaneous signals, as though the keys were being affected in concert.[21]

Most intriguing, in much of the research on psychokinesis, mental influence of any variety had produced measurable effects, no matter how far the distance between the sender or what point in time he generated his intention. According to the experimental evidence, the power of thought transcended time and space.

By the time these revisionists were finished, they had torn up the rule book and scattered it to the four winds. Mind in some way appeared to be inextricably connected to matter and, indeed, was capable of altering it. Physical matter could be influenced, even irrevocably altered, not simply by force, but through the simple act of formulating a thought.

Nevertheless, the evidence from these frontier scientists left three fundamental questions unanswered. Through what physical mechanisms do thoughts affect reality? At the time of this writing, some highly publicized studies of mass prayer showed no effect. Were certain conditions and preparatory states of mind more conducive to success than others? How much power did a thought have, for good or ill? How much of our lives could a thought actually change?

Most of the initial discoveries about consciousness occurred more than 30 years ago. More recent discoveries in frontier quantum physics and in laboratories around the globe offer answers to some of those questions. They provide evidence that our world is highly malleable, open to constant subtle influence. Recent research demonstrates that living things are constant transmitters and receivers of measurable energy. New models of consciousness portray it as an entity capable of trespassing physical boundaries of every description.

Intention appears to be something akin to a tuning fork, causing the tuning forks of other things in the universe to resonate at the same frequency.

The latest studies of the effect of mind on matter suggest that intention has variable effects that depend on the state of the host, and the time and the place where it originates. Intention has already been employed in many quarters to cure illness, alter physical processes and influence events. It is not a special gift but a learned skill, readily taught. Indeed, we already use intention in many aspects of our daily lives.

A body of research also suggests that the power of an intention multiplies, depending upon how many people are thinking the same thought at the same time.[22]

The Intention Experiment consists of three aspects. The main body of the book (chapters 1–12) attempts to synthesize all the experimental evidence that exists on intention into a coherent scientific theory of how intention works, how it can be used in your life and which conditions optimize its effect.

The second portion of the book (chapter 13) offers a blueprint for using intention effectively in your own life through a series of exercises and recommendations for how best to 'power up'. This portion is also an exercise in frontier science. I am not an expert in human potential, so this is not a self-help manual, but a journey of discovery for me as well as you. I have extrapolated this programme from scientific evidence describing those circumstances that created the most positive results in psychokinetic laboratory experiences. We know for certain that these techniques have generated success under controlled experimental laboratory conditions, but I cannot guarantee they will work in your life. By making use of them, you will, in effect, engage in an ongoing personal experiment.

The final section of the book (chapters 14 and 15) consists of a series of personal and group experiments. Chapter 14 outlines a series of informal experiments on the use of intention in your own life for you to carry out individually. These mini 'experiments' are also intended to be pieces of research. You will have the opportunity to post your results on our website and share them with other readers.

Besides these individual experiments, I have also designed a series of large group experiments to be carried out by the readers of this book (chapter 15). With the aid of our highly experienced scientific team, *The Intention Experiment* will conduct periodic large-scale experiments to determine whether the focused intention of its readers has an effect on scientifically quantifiable targets.

All it requires is that you read the book, digest its contents, log on to the website (www.theintentionexperiment.com) and, after following the instructions and exercises at the back of this book, send out some highly specific thoughts, as and when described on the site. The first such studies will be carried out by the German physicist Fritz-Albert Popp, vice-president of the International Institute of Biophysics in Neuss, Germany (www.lifescientists.de), and his team of seven, psychologist Gary Schwartz and his colleagues at the University of Arizona at Tucson, and Marilyn Schlitz and Dean Radin of the Institute of Noetic Sciences.

Website experts have collaborated with our scientific team to design log-on protocols to enable us to identify which characteristics of a group or aspects of their thoughts produce the most effective results. For each intention experiment, a target will be selected – a specific living thing or a population where change caused by group intention can be measured. We have started with algae, the lowliest of subjects (see chapter 12), and, with every experiment, we will move on to an increasingly complex living target.

Our plans are ambitious: to tackle a number of societal ills. One eventual human target might be patients with a wound. It is known and accepted that wounds generally heal at a particular, quantifiable rate with a precise pattern.[23] Any departure from the norm can be precisely measured and shown to be an experimental effect. In that instance, our aim would be to determine whether focused group intention will enable wounds to heal more quickly than usual.

Naturally, you don't have to participate in our experiments. If you don't wish to get involved, you can read about the intention experiments of others, and use some of that information to inform how you use intention in your life.

Please do not casually participate in the experiments. In order for the experiment to work properly, you must read the book and digest its contents fully beforehand. The experimental evidence suggests that those who are the most effective have trained their minds, much as athletes train their muscles, to maximize their chances of success.

In order to discourage uncommitted participation, *The Intention Experiment* website contains a complicated password comprising some words or ideas from the book (which will change slightly every few months). In order to be part of the experiment, you will have to log on with the password and you will have to have read the book and understood it.

The website (www.theintentionexperiment.com) has a running clock (set to US Eastern Standard Time and Greenwich Mean Time). At a particular moment on a date specified on the website, you will be asked to send a carefully worded, detailed intention, depending on the target site.

Once finished, the results of the experiments will be analysed and data-crunched by our scientific team, examined by a neutral statistician, and then published on the website and in subsequent printings of this book. The website will thus become the living sequel to the book you are holding in your hands. You simply need to consult the website periodically for announcements of the date of every experiment.

Hundreds of well-designed studies of group intention and remote mental influence have demonstrated significant results. Nevertheless, it might be the case that our experiments will not produce demonstrable, measurable effects, at first or indeed ever. As reputable scientists and objective researchers, we are duty-bound to report the data we have. As with all science, failure is instructive, helping us to refine the design of the experiments and the premises that they are based upon.

As you read this book, keep in mind that this is a work of frontier science. Science is a relentless process of self-correction. Assumptions originally considered as fact must often ultimately be discarded. Many – indeed, most – of the conclusions drawn in this book are bound to be amended or refined at a later date.

By reading this book and participating in its experiments you may well contribute to the world's knowledge, and possibly further a paradigm shift in our understanding of how the world works. Indeed, the power of mass intention may ultimately be the force that shifts the tide towards repair and renewal of the planet. When combined with hundreds of thousands of others, your solitary voice, now one barely audible note, could transmute into a thunderous symphony.

My own motive for writing *The Intention Experiment* was to make a statement about the extraordinary nature and power of consciousness. It may prove true that a single collective, directed thought is all it takes to change the world.

PART ONE

The Science of Intention

A human being is part of the whole, called by us
'universe', a part limited in time and space. He
experiences himself, his thoughts and feelings as
something separated from the rest – a kind of
optical delusion of his consciousness.

Albert Einstein

CHAPTER ONE
Mutable Matter

FEW PLACES IN THE GALAXY are as cold as the helium-dilution refrigerator in Tom Rosenbaum's lab. Temperatures in the refrigerator – a boiler-sized circular apparatus with a number of cylinders – can descend to a few thousandths of a degree above absolute zero, almost 273°C below freezing – three thousand times colder than the farthest reaches of outer space. For two days, liquid nitrogen and helium circulate around the refrigerator, and then three pumps constantly blasting out gaseous helium take the temperature down to the final rung. Without heat of any description, the atoms in matter slow to a crawl. At this scale of coldness, the universe would grind to a halt. It is the scientific equivalent of hell freezing over.

Absolute zero is the preferred temperature of a physicist like Tom Rosenbaum. At 47, as a distinguished professor of physics at the University of Chicago and former head of the James Franck Institute, Rosenbaum was in the vanguard of experimental physicists who liked exploring the limits of disorder in condensed-matter physics, the study of the inner workings of liquids and solids when their underlying order was disturbed.[1] In physics, if you want to find out how something behaves, the best way is simply to make it uncomfortable and then see what happens. Creating disorder usually involves adding heat or applying a magnetic field to determine how it will react when disturbed and also to determine which spin position – or magnetic orientation – the atoms will choose.

Most of his colleagues in condensed-matter physics remained interested in symmetrical systems such as crystalline solids, whose atoms are arranged in orderly array, like eggs in a carton, but Rosenbaum was drawn to strange systems that were inherently disordered – to which more conventional quantum physicists referred disparagingly as 'dirt'. In dirt, he believed, lay exposed the unprobed secrets of the quantum universe, uncharted territory that he was happy to navigate. He loved the challenge posed by spin glasses, strange hybrids of crystals, with magnetic properties, technically considered slow-moving liquids. Unlike a crystal, whose atoms point in the same direction in perfect alignment, the tiny magnets associated with the atoms of a spin glass are wayward and frozen in disarray.

The use of extreme coldness allowed Rosenbaum to slow down the atoms of these strange compounds enough to observe them minutely, and to tease out their quantum mechanical essence. At temperatures near to absolute zero, when their atoms are nearly stationary, they begin taking on new collective properties. Rosenbaum was fascinated by the recent discovery that systems disorderly at room temperature display a conformist streak once they are cooled down. For once, these delinquent atoms begin to act in concert.

Examining how molecules behave as a group in various circumstances is highly instructive about the essential nature of matter. In my own journey of discovery, Rosenbaum's laboratory seemed the most appropriate place to begin. There, at those lowest temperatures where everything occurs in slow motion, the true nature of the most basic constituents of the universe might be revealed. I was looking for evidence of ways in which the components of our physical universe, which we think of as fully realized, are capable of being fundamentally altered. I also wondered whether it could be shown that quantum behaviour like the observer effect occurs outside the subatomic world, in the world of the everyday. What Rosenbaum had discovered in his refrigerator might offer some vital clues as to how every object or organism in the physical world, which classical physics depicts as an irreversible fact, a finalized assemblage only changeable by the brute force of Newtonian physics, could be affected and ultimately altered by the energy of a thought.

According to the second law of thermodynamics, all physical processes in the universe can only flow from a state of greater to lesser energy. We throw a stone into a river and the ripple it makes eventually stops. A cup of hot coffee left standing can only grow cold. Things inevitably fall apart; everything travels in a single direction, from order to disorder.

But this might not always be inevitable, Rosenbaum believed. Recent discoveries about disordered systems suggested that certain materials, under certain circumstances, might counteract the laws of entropy and come together rather than fall apart. Was it possible that matter could go in the opposite direction, from disorder to greater order?

For ten years Rosenbaum and his students at the James Franck Institute had been asking that question of a small chunk of lithium holmium fluoride salt. Inside Rosenbaum's refrigerator lay a perfect chip of rose-coloured crystal, no bigger than the head of a pencil, wrapped in two sets of copper coils. Over the years, after many experiments with spin glasses, Rosenbaum had grown very fond of these dazzling little specimens, one of the most naturally magnetic substances on earth. This characteristic presented the perfect situation in which to study disorder, but only after he had altered the crystal beyond recognition into a disordered substance.

He had first instructed the laboratory that grew the crystals to combine the holmium with fluorine and lithium, the first metal on the periodic table. The resulting lithium holmium fluoride salt was compliant and predictable – a highly ordered substance whose atoms behaved like a sea of microscopic compasses all pointing north. Rosenbaum then had wreaked havoc on the original salt compound, instructing the lab to rip out a number of the atoms of holmium, bit by bit, and replace them with yttrium, a silvery metal without such natural magnetic attraction, until he was left with a strange hybrid of a compound: a salt called lithium holmium yttrium tetrafluoride.

By virtually eliminating the magnetic properties of the compound, Rosenbaum eventually had created spin-glass anarchy – the atoms of this Frankenstein monstrosity pointing any way they liked. Being able to manipulate the essential property of elements like holmium by creating weird new compounds so cavalierly was a little like having ultimate control over matter itself. With these new spin-glass compounds, Rosenbaum could virtually change the properties of the compound at will; he could make the atoms orientate in a particular direction, or freeze them in some random pattern.

Nevertheless, his omnipotence had a limit. Rosenbaum's holmium compounds behaved themselves in some regards, but not in others. One thing he could not do was to get them to obey the laws of temperature. No matter how cold Rosenbaum made his refrigerator, the atoms inside them resisted any sort of ordered orientation, like an army refusing to march in step. If Rosenbaum was playing God with his spin glasses, the crystal was Adam, stubbornly refusing to obey His most fundamental law.

Sharing Rosenbaum's curiosity about the strange property of the crystal compound was a young student called Sayantani Ghosh, one of his star PhD candidates. Sai, as her friends called her, a native of India, had graduated with a first-class honours degree from Cambridge, after which she had chosen Tom's lab for her doctoral programme in 1999. Almost immediately, she had distinguished herself by winning the Gregor Wentzel Prize, given each year by the University of Chicago's physics department to the best first-year graduate student teaching assistant. The slight 23-year-old, who at first glance appeared abashed, hiding behind her copious dark hair, had soon impressed her peers and teachers alike with her bold authority, a rarity among science students, and her ability to translate complex ideas to the level an undergraduate could comprehend. Sai shared the distinction of winning the coveted prize with only one other woman since its inception 25 years before.

According to the laws of classical physics, applying a magnetic field will disrupt the magnetic alignment of a substance's atoms. The degree to which this happens is the salt's 'magnetic susceptibility'. The usual pattern with a disordered substance is that it will respond to the magnetic field for a time and then plateau and tail off, as the temperature drops or the magnetic field reaches a point of magnetic saturation. The atoms will no longer be able to flip in the same direction as that of the magnetic field and so will begin to slow down.

In Sai's first experiments, the atoms in the lithium holmium yttrium salt, as predicted, grew wildly excited with the application of the magnetic field. But then, as Sai increased the field, something strange began to happen. The more she turned up the frequency, the faster the atoms continued to flip over. What is more, all the atoms, which had been in a state of disarray, began pointing in the same direction and operating as a collective whole. Then, small clusters of about 260 atoms aligned, forming 'oscillators', spinning collectively in one direction or another. No matter how strong the magnetic field that Sai applied, the atoms remained stubbornly aligned with each other, acting in concert. This self-organization persisted for 10 seconds.

At first, Sai and Rosenbaum thought these effects might have something to do with the strange effects of the remaining atoms of holmium, known to be one of the very few substances in the world with such long-range internal forces that in some quarters it was described and worked out mathematically as something existing in another dimension.[2] Although they didn't understand the phenomenon they had observed, they wrote up their results, which were published in the journal *Science* in 2002.[3]

Rosenbaum decided to carry out another experiment to attempt to isolate the property in the crystal's essential nature that had enabled it to override such strong outside influences. He left the study's design to his bright young graduate student, suggesting only that she create a computerized three-dimensional mathematical simulation of the experiment she had intended to carry out. In experiments of this nature on such tiny matter, physicists must rely on a computerized simulation to confirm mathematically the reactions they are witnessing experimentally.

Sai spent months developing the computer code and building her simulation. The plan was to find out a bit more about the salt's magnetic capability, by applying two systems of disorder to the crystal chip: higher temperatures and a stronger magnetic field.

She prepared the sample by placing it in a little 2.4 x 4.8 cm copper holder, then wrapped two coils around the tiny crystal: one a gradiometer, to measure its magnetic susceptibility and the direction of spin of the individual atoms, and the other to cancel out any random flux affecting the atoms inside.

A connection attached to her PC would enable her to change the voltage, the magnetic field or the temperature, and would record any changes whenever she altered one of the variables by the tiniest degree.

She began lowering the temperature, a fraction of a kelvin (K) at a time, and then began applying a stronger magnetic field. To her amazement, the atoms kept aligning progressively. Then she tried applying heat, and discovered they again aligned. No matter what she did, in every instance the atoms ignored the outside interference. Although she and Tom had flushed out most of the compound's magnetic component, of its own volition, as it were, it was turning into a larger and larger magnet.

That's weird, she thought. Perhaps she should take more data, just to ensure they had encountered nothing strange in the system.

She repeated her experiment over six months until the early spring of 2002, when her computer simulation was finally complete. One evening, she mapped the results of the simulation on a graph, and then she superimposed the results from her actual experiment. It was as though she had drawn a single line. There on the computer screen was a perfect duplicate: the diagonal line formed from the computer simulation lay exactly over the diagonal line created from the results of the experiment itself. What she had witnessed in the little crystal was not an artefact, but something real that she had now reproduced in her computer simulation. She had even mapped out where the atoms should have been on the graph, had they been obeying the usual laws of physics. But there they were in a line: a law completely unto themselves.

She wrote Rosenbaum a guarded email late that evening: 'I've got something interesting to show you in the morning.' The following day, they examined her graph. There was no other possibility, they both realized; the atoms had been ignoring her and instead were controlled by the activity of their neighbours. No matter whether she blasted the crystal with a strong magnetic field or an increase in temperature, the atoms overrode this outside disturbance.

The only explanation was that the atoms in the sample crystal were internally organizing and behaving like one single giant atom. All the atoms, they realized with some alarm, must be entangled.

One of the strangest aspects of quantum physics is a feature called 'non-locality', also poetically referred to as 'quantum entanglement'. The Danish physicist Niels Bohr discovered that once subatomic particles such as electrons or photons are in contact, they remain cognizant of and influenced by each other instantaneously over any distance forever, despite the absence of the usual things that physicists understand are responsible for influence, such as an exchange of force or energy. When entangled, the actions – for instance, the magnetic orientation – of one will always influence the other in the same or the opposite direction, no matter how far they are separated. Erwin Schrödinger, another one of the original architects of quantum theory, believed that the discovery of non-locality represented no less than quantum theory's defining moment – its central property and premise.

The activity of entangled particles is analogous to a set of twins being separated at birth, but retaining identical interests and a telepathic connection forever. One lives in Colorado, and the other in London. Although they never meet again, both like the colour blue. Both take a job in engineering. Both like to ski; in fact when one falls down and breaks his right leg at Vale, his twin breaks his right leg at precisely that moment, even though he is 4000 miles away, sipping a latte at Starbucks.[4] Albert Einstein refused to accept non-locality, referring to it disparagingly as '*spukhafte Fernwirkungen*' or 'spooky action at a distance'. This type of instantaneous connection would require information travelling faster than the speed of light, he argued through a famous thought experiment, which would violate his own special relativity theory.[5] Since the formulation of Einstein's theory, the speed of light (299,792,458 metres per second) has been used as the absolute limiting factor on how quickly one thing can affect something else. Things are not supposed to be able to affect other things faster than the time it would take the first thing to travel to the second thing at the speed of light.

Nevertheless, modern physicists, such as Alain Aspect and his colleagues in Paris, have demonstrated decisively that the speed of light is not an absolute outer boundary in the subatomic world. Aspect's experiment, which concerned two photons fired off from a single atom, showed that the measurement of one photon instantaneously affected the position of the second photon[6] so that it has the same or opposite spin or position (as IBM physicist Charles H. Bennett once put it, 'opposite luck').[7] The two photons continued to talk to each other and whatever happened to one was identical to, or the very opposite of, what happened to the other. Today, even the most conservative physicists accept non-locality as a strange feature of subatomic reality.[8]

Most quantum experiments incorporate some test of Bell's Inequality. This famous experiment in quantum physics was carried out by John Bell, an Irish physicist who developed a practical means to test how quantum particles really behaved.[9] This simple test required that you get two quantum particles that had once been in contact, separate them and then take measurements of the two. It is analogous to a couple named Daphne and Ted who have once been together but are now separated. Daphne can choose one of two possible directions to go in and so can Ted. According to our common-sense view of reality, Daphne's choice should be utterly independent of Ted's.

When Bell carried out his experiment, the expectation was that one of the measurements would be larger than the other – a demonstration of 'inequality'. However, a comparison of the measurements showed that both were the same and so his inequality was 'violated'. Some invisible wire appeared to be connecting these quantum particles across space, to make them follow each other. Ever since, physicists have understood that when a violation of Bell's Inequality occurs, it means that two things are entangled.

Bell's Inequality has enormous implications for our understanding of the universe. By accepting non-locality as a natural facet of nature we are acknowledging that two of the bedrocks on which our world view rests are wrong: that influence only occurs over time and distance, and that particles like Daphne and Ted, and indeed the things that are made up of particles, only exist independently of each other.

Although modern physicists now accept non-locality as a given feature of the quantum world, they console themselves by maintaining that this strange, counter-intuitive property of the subatomic universe does not apply to anything bigger than a photon or electron. Once things got to the level of atoms and molecules, which in the world of physics is considered 'macroscopic', or large, the universe started behaving itself again, according to predictable, measurable, Newtonian laws.

With one tiny thumbnail's worth of crystal, Rosenbaum and his graduate student demolished that delineation. They had demonstrated that big things like atoms were non-locally connected, even in matter so large you could hold it in your hand. Never before had quantum non-locality been demonstrated on such a scale. Although the specimen had been only a tiny chip of salt, to the subatomic particle, it was a palatial country mansion, housing a billion billion $(1,000,000,000,000,000,000$ or $10^{18})$ atoms. Rosenbaum, ordinarily loathe to speculate about what he could not yet explain, realized that they had uncovered something extraordinary about the nature of the universe. And I realized they had discovered a mechanism for intention: they had demonstrated that atoms, the essential constituents of matter, could be affected by non-local influence. Large things like crystals were not playing by the grand rules of the game, but by the anarchic rules of the quantum world, maintaining invisible connections without obvious cause.

In 2002, after Sai wrote up their findings, Rosenbaum polished up the wording and sent off their paper to *Nature*, a journal notorious for conservatism and exacting peer review. After four months of responding to the suggestions of reviewers, Ghosh finally got her paper published in the world's premier scientific journal, a laudatory feat for a 26-year-old graduate student.[10]

One of the reviewers, Vlatko Vedral, noted the experiment with a mix of interest and frustration.[11] A Yugoslav who had studied at Imperial College, London, during his country's civil war and subsequent collapse, Vedral had distinguished himself in his adopted country and been chosen to head up quantum information science at the University of Leeds. Vedral, who was tall and leonine, was part of a small group in Vienna working on frontier quantum physics, including entanglement.

Vedral first theoretically predicted the effect that Ghosh and Rosenbaum eventually found three years later. He had submitted the paper to *Nature* in 2001, but the journal, which preferred experiment to theory, had rejected it. Eventually, Vedral managed to publish his paper in *Physical Review Letters*, the premier physics journal.[12] After *Nature* decided to publish Ghosh's study, its editors threw him a conciliatory bone. They allowed him to be a reviewer on the paper, and then offered him a place in the same issue to write an opinion piece on the findings.

In the article, Vedral allowed himself some speculation. Quantum physics is accepted as the most accurate means of describing how atoms combine to form molecules, he wrote, and since molecular relationship is the basis of all chemistry, and chemistry is the basis of biology, the magic of entanglement could well be the key to life itself.[13]

Vedral and a number of others in his circle did not believe that this effect was unique to holmium. The central problem in uncovering entanglement is the primitive state of our technology; isolating and observing this effect is only possible at the moment by slowing atoms down so much in such cold conditions that they are hardly moving. Nevertheless, a number of physicists had observed entanglement in matter at 200 K, or $-73°C$ – a temperature that can be found on Earth in some of its very coldest places.

Other researchers have proved mathematically that everywhere, even inside of our own bodies, atoms and molecules are engaged in an instantaneous and ceaseless passing back and forth of information. Thomas Durt of Vrije University in Brussels demonstrated through elegant mathematical formulations that almost all quantum interactions produce entanglement, no matter what the internal or surrounding conditions. Even photons, the tiniest particles of light emanating from stars, are entangled with every atom they meet on their way to earth.[14] Entanglement at normal temperatures appears to be a natural condition of the universe, even in our bodies. Every interaction between every electron inside of us creates entanglement. According to Benni Reznik, a theoretical physicist at Tel Aviv University in Israel, even the empty space around us is heaving with entangled particles.[15]

The English mathematician Paul Dirac, an architect of quantum field theory, first postulated that there is no such thing as nothingness, or empty space. Even if you tipped all matter and energy out of the universe and examined all the 'empty' space between the stars you would discover a netherworld world teeming with subatomic activity.

In the world of classical physics, a field is a region of influence, in which two or more points are connected by a force, like gravity or electromagnetism. However, in the world of the quantum particle, fields are created by exchanges of energy. According to Heisenberg's uncertainty principle, one reason that quantum particles are ultimately unknowable is because their energy is always being redistributed in a dynamic pattern. Although often rendered as tiny billiard balls, subatomic particles more closely resemble little packets of vibrating waves, passing energy back and forth as if in an endless game of basketball. All elementary particles interact with each other by exchanging energy through what are considered temporary or 'virtual' quantum particles. These are believed to appear out of nowhere, combining and annihilating each other in less than an instant, causing random fluctuations of energy without any apparent cause. Virtual particles, or negative energy states, do not take physical form, so we cannot actually observe them. Even 'real' particles are nothing more than a little knot of energy, which briefly emerge and disappear back into the underlying energy field.

These back-and-forth passes, which rise to an extraordinarily large ground state of energy, are known collectively as the Zero Point Field. The field is called 'zero point' because even at temperatures of absolute zero, when all matter theoretically should stop moving, these tiny fluctuations are still detectable. Even at the coldest place in the universe, subatomic matter never comes to rest, but carries on this little energy tango.[16]

The energy generated by every one of these exchanges between particles is unimaginably tiny – about half a photon's worth. However, if all exchanges between all subatomic particles in the universe were to be added up, it would produce an inexhaustible supply of energy of unfathomable proportions, exceeding all energy in matter by a factor of 10^{40}, or 1 followed by 40 zeros.[17] Richard Feynman himself once remarked that the energy in a cubic metre of space was enough to boil all the oceans of the world.[18]

After the discoveries of Heisenberg about Zero Point energy, most conventional physicists have subtracted the figures symbolizing Zero Point energy from their equations. They assumed that, because the Zero Point Field was ever present in matter, it did not change anything and so could be safely 'renormalized' away. However, in 1973, when trying to work out an alternative to fossil fuel during the petrol crisis, American physicist Hal Puthoff, inspired by the Russian Andrei Sakharov, began trying to figure out how to harness the teeming energy of empty space for transport on earth and to distant galaxies. Puthoff spent more than 30 years examining the Zero Point Field. With some colleagues, he had proved that this constant energy exchange of all subatomic matter with the Zero Point Field accounts for the stability of the hydrogen atom, and, by implication, the stability of all matter.[19] Remove the Zero Point Field and all matter would collapse in on itself. He also demonstrated that Zero Point energy is responsible for two basic properties of mass: inertia and gravity.[20]

Puthoff also worked on a multimillion-dollar project funded by Lockheed Martin and a variety of American universities, to develop Zero Point energy for space travel – a programme that finally went public in 2006.

Many strange properties of the quantum world, like uncertainty or entanglement, could be explained if you factored in the constant interaction of all quantum particles with the Zero Point Field. To Puthoff, science's understanding of the nature of entanglement was analogous to two sticks stuck in the sand at the edge of the ocean, about to be hit by a huge wave. If they both were knocked over, and you did not know about the wave, you would think that one stick was affecting the other and call it a non-local effect. The constant inter-action of quantum particles with the Zero Point Field might be the underlying mechanism for non-local effects between particles, allow-ing one particle to be in touch with every other particle at any moment.[21]

Benni Reznik's work in Israel with the Zero Point Field and entan-glement began mathematically with a central question: what would happen to a hypothetical pair of probes interacting with the Zero Point Field? According to his calculations, once they began interact-ing with the Zero Point Field, the probes would begin talking to each other and ultimately become entangled.[22]

If all matter in the universe were interacting with the Zero Point Field, it meant, quite simply, that all matter was interconnected and potentially entangled throughout the cosmos through quantum waves.[23] And if we and all of empty space are a mass of entanglement, we must be establishing invisible connections with things at a distance to ourselves. Acknowledging the existence of the Zero Point Field and entanglement offers a ready mechanism for why signals being generated by the power of thought can be picked up by some-one else many miles away.

* * *

Sai Ghosh had proved that non-locality existed in the large building blocks of matter and the other scientists proved that all matter in the universe was, in a sense, a satellite of a large central energy field. But how could matter be affected by this connection? The central assumption of all of classical physics is that large material things in the universe are set pieces, a *fait accompli* of manufacture. How can they possibly be changed?

Vedral had an opportunity to examine this question when he was invited to work with the renowned quantum physicist Anton Zeilinger. Zeilinger's Institute for Experimental Physics lab at the University of Vienna was at the very frontier of some of the most exotic research into the nature of quantum properties. Zeilinger himself was profoundly dissatisfied with the current scientific explanation of nature, and he had passed on that dissatisfaction and the quest to resolve it to his students.

In a flamboyant gesture, Zeilinger and his team had entangled a pair of photons from beneath the River Danube. They had set up a quantum channel via a glass fibre and run it across the river bed of the Danube. In his lab, Zeilinger liked to refer to individual photons as Alice and Bob, and sometimes, if he needed a third photon, Carol or Charlie. Alice and Bob, separated by 600 metres of river and nowhere in sight of each other, maintained a non-local connection.[24]

Zeilinger was particularly interested in superposition, and the implications of the Copenhagen Interpretation – that subatomic particles exist only in a state of potential. Could objects, and not simply the subatomic particles that compose them, he wondered, exist in this hall-of-mirrors state? To test this question, Zeilinger employed a piece of equipment called a Talbot Lau interferometer, developed by some colleagues at MIT, using a variation on the famous double-slit experiment of Thomas Young, a British physicist of the nineteenth century. In Young's experiment, a beam of pure light is sent through a single hole, or slit, in a piece of cardboard, then passes through a second screen with two holes before finally arriving at a third, blank screen.

When two waves are in phase (that is, peaking and troughing at the same time), and bump into each other – technically called 'interference' – the combined intensity of the waves is greater than each individual amplitude. The signal gets stronger. This amounts to an imprinting or exchange of information, called 'constructive interference'. If one is peaking when the other troughs, they tend to cancel each other out – called 'destructive interference'. With constructive interference, when all the waves are wiggling in synch, the light will get brighter; destructive interference will cancel out the light and result in complete darkness.

In the experiment, the light passing through the two holes forms a zebra pattern of alternating dark and light bands on the final blank screen. If light were simply a series of particles, two of the brightest patches would appear directly behind the two holes of the second screen. However, the brightest portion of the pattern is halfway between the two holes, caused by the combined amplitude of those waves that most interfere with each other. From this pattern, Young was the first to realize that light beaming through the two holes spreads out in overlapping waves.

A modern variation of the experiment fires off single photons through the double slit. These single photons also produce zebra patterns on the screen, demonstrating that even single units of light travel as a smeared-out wave with a large sphere of influence.

Twentieth-century physicists went on to use Young's experiment with other individual quantum particles, and held it up as proof that quantum physics had Through-the-Looking-Glass properties: *quantum entities acted wavelike and travelled though both slits at once.* Fire a stream of electrons at the triple screens, and you end up with the interference patterns of alternating light and dark patches, just as you do with a beam of light. Since you need at least two waves to create such interference patterns, the implication of the experiment is that the photon is somehow mysteriously able to travel through both slits at the same time and interfere with itself when it reunites.

The double-slit experiment encapsulates the central mystery of quantum physics – the idea that a subatomic particle is not a single seat but the entire stadium. It also demonstrates the principle that electrons, which exist in a hermetic quantum state, are ultimately unknowable. You could not identify something about a quantum entity without stopping the particle in its tracks, at which point it would collapse to a single point.

In Zeilinger's adaptation of the slit experiment, using molecules instead of subatomic particles, the interferometer contained an array of slits in the first screen, and a grating of identical parallel slits in the second one, whose purpose was to diffract (or deflect) the molecules passing by. The third grating, turned perpendicular to the beam of molecules, acted as a scanning 'mask', with the ability to calculate the size of the waves of any of the molecules passing through, by means of a highly sensitive laser detector to locate the positions of the molecules and their interference patterns.

For the initial experiment, Zeilinger and his team carefully chose a batch of fullerene molecules, or 'buckyballs' made of 60 carbon atoms. At one nanometre apiece, these are the behemoths of the molecular world. They selected fullerene not only for its size but also for its neat arrangement, with a shape like a tiny symmetrical football.

It was a delicate operation. Zeilinger's group had to work with just the right temperature; heating the molecules just a hair too much would cause them to disintegrate. Zeilinger heated the fullerenes to 900 K so they would create an intense molecular beam, then fired them through the first screen; they then passed through the second screen before making a pattern on the final screen. The results were unequivocal. Each molecule displayed the ability to create interference patterns with itself. Some of the largest units of physical matter had not 'localized' into their final state. Like a subatomic particle, these giant molecules had not yet gelled into anything real.

The Vienna team scouted out some other molecules that were double the size and oddly shaped to see if geometrically asymmetric molecules also demonstrated the same magical properties. They settled on gigantic fluorinated American football-shaped molecules of 70 carbon atoms and pancake-shaped tetraphenylporphyrin, a derivative of the biodye present in chlorophyll. At more than 100 atoms apiece, both of these entities are among the largest molecules on the planet. Again, each one created an interference pattern with itself.

Zeilinger's group repeatedly demonstrated that the molecules could be two places at once, which meant that they remained in a state of superposition even at this large scale.[25] They had proved the unthinkable: the largest components of physical matter and living things exist in a malleable state.[26]

Sai Ghosh didn't often think about the implications of her discovery. She was content with the knowledge that her experiment had made a very nice paper, and might help along her career as an assistant professor involved in research into miniaturization, the direction she believed quantum mechanics was heading. Occasionally, she allowed herself to speculate that her crystal might have proved something important about the nature of the universe. But she was only a postgraduate student. What did she, after all, really know about how the world worked?

But to me, Ghosh's research and Zeilinger's work on the double-slit experiment represent two defining moments in modern physics. Ghosh's experiments show that an invisible connection exists between the fundamental elements of matter, which is often so strong that it can override classical methods of influence, such as heat or a push. Zeilinger's work demonstrated something even more astonishing. Large matter was neither something solid and stable nor something that necessarily behaved according to Newtonian rules. Molecules needed some other influence to settle them into a completed state of being.

Theirs were the first evidence that the peculiar properties of quantum physics do not simply occur at the quantum level with subatomic particles, but also in the world of visible matter. Molecules also exist in a state of pure potential, not a final actuality. Under certain circumstances, they escape Newtonian rules of force and display quantum non-local effects. The fact that something as large as a molecule can become entangled suggests that there are not two rule books – the physics of the large and the physics of the small – but only a single rule book for all of life.

These two experiments also hold the key to a science of intention – how thoughts are able to affect finished, solid matter. They suggest that the observer effect occurs not simply in the world of the quantum particle but also in the world of the everyday. Things no longer should be seen to exist in and of themselves but, like a quantum particle, only in relationship. Co-creation and influence may be a basic, inherent property of life. Our observation of every component in our world may help to determine its final state, which suggests that we are likely to be influencing every large thing we see around us. When we enter a crowded room, when we engage with our partners and our children, when we gaze up at the sky, we may be creating and even influencing at every moment. We can't yet demonstrate this at normal temperatures; our equipment is still too crude. But we already have some preliminary proof: the physical world – matter itself – appears to be malleable, susceptible to influence from the outside.

CHAPTER TWO

The Human Antenna

IN 1951, AT THE AGE OF SEVEN, Gary Schwartz made a remark-able discovery. He had been trying to get a good picture on the family's television set. The recently acquired black and white Magnavox set encased behind the doors of its boxed walnut console fascinated him, not because of the people in the moving pictures so much as the means by which they arrived in his living room in the first place. The mechanisms of the relatively new invention remained a mystery, even to most adults. Television, like any other electrical gadget, was something the precocious child longed to take apart and understand. This passion had already found expression with the worn-out radios given to him by his grandfather. Ignatz Schwartz sold replacement tubes for televisions and radios in his drug store in Great Neck, Long Island, and those that were beyond repair were handed over to his grandson to disassemble. In a corner of Gary's bedroom lay a mass of experimental debris – tubes, resistors and the carcasses of radios heaped on the cosmetic display racks he had borrowed from his grandfather – the first signs of what would become a lifelong fascination with electronics.

Gary knew that the way you twisted the rabbit-ear antenna on top of the television would determine the clarity of the picture. His father had explained that television sets were powered by something invisible, similar to radio waves, that flew through the air and were somehow translated into an image. Gary had even carried out some rudimentary experiments. When you stood somewhere between the antenna and the television, you could make the picture go away. When you touched the antenna in certain ways, you made the picture clearer.

One day, on a whim, Gary unscrewed the antenna and placed his finger on the screw where the cable had been. What had been a mass of squiggles and static noise on the screen suddenly coalesced into a perfect image. Even at that young age, he had understood that he had witnessed something extraordinary about human beings: his body was acting like a television antenna, a receiver of this invisible information. He tried the same experiment with a radio – substituting his finger for the antenna, and the same thing happened. Something in the makeup of a person was not unlike the rabbit ears that helped produce his television image. He too was a *receiver* of invisible information, with the ability to pick up signals transmitted across time and space.

Until he was 15, however, he could not visualize what these signals were made of. He had learned to play the electric guitar and had often wondered what unseen influences allowed the instrument to create different sounds. He could play the same note, middle C, and yet produce more of a treble or bass sound, depending on which way he turned the knob. How was it possible that a single note could sound so different? For a science project, he created multiple-track recordings of his music and then located a company in upstate New York that had equipment designed to analyse the frequency of sound. When he fed his recordings into the equipment, it quickly deconstructed the notes down to their essence. Each note registered as a batch of squiggles across the screen of the cathode-ray tube in front of him – a complex mix of hundreds of frequencies representing a blend of overtones that would subtly change when he turned the knob to treble or bass. He knew that these frequencies were waves, represented on the monitor as a sideways S, or sine curve, like a skipping rope held at both ends and wriggled, and that they had periodic oscillations, or fluctuations, similar to the waves on Long Island Sound. Every time he spoke, he knew he generated similar frequencies through his voice. He remembered his early television experiments and wondered whether a field of energy pulsated inside him and shared a kinship with sound waves.[1]

Gary's childhood experiments may have been rudimentary, but he had already stumbled across the central mechanism of intention. Something in the quality of our thoughts was a constant transmission, not unlike a television station.

As an adult, Schwartz, still a bustling dynamo of enthusiasms, found an outlet in psychophysiology, then a fledgling study of the effect of the mind on the body. By the time he had accepted a post at the University of Arizona, which was known for encouraging freedom of research among its faculty, he had grown fascinated by biofeedback and the ways in which the mind could control blood pressure and a variety of illnesses – and the powerful physical effect of different types of thoughts.[2]

One weekend in 1994, at a conference on the relationship between love and energy, he sat in on a lecture by physicist Elmer Green, one of the pioneers of biofeedback. Green, like Schwartz, had grown interested in the energy being transmitted by the mind. To examine this more closely, he had decided to study remote healers and to determine whether they sent out more electrical energy than usual while in the process of healing.

Green reported in his lecture that he had built a room whose four walls and ceiling were entirely made of copper, and were attached to microvolt electroencephalogram (EEG) amplifiers – the kind used to measure the electrical activity in the brain. Ordinarily, an EEG amplifier is attached to a cap with imbedded electrodes, each of which records separate electrical discharges from different places in the brain. The cap is placed on a person's head, and the electrical activity picked up is displayed on the amplifier. EEG amplifiers are extraordinarily sensitive, capable of picking up the most minute of effects – even one-millionth of a volt of electricity.

In remote healing, Green suspected that the signal produced was electrical and emanated from the healer's hands. The copper wall acted like a giant antenna, magnifying the ability to detect the electricity from the healers and enabling Green to capture it from five directions.

He discovered that, whenever a healer sent healing, the EEG amplifier often recorded it as a huge surge of electrostatic charge, the same kind of the build up and discharge of electrons that occurs after you shuffle your feet along a new carpet and then touch a metal doorknob.[3]

In the early days of the copper wall experiment, Green had been faced with an enormous problem. Whenever a healer so much as wriggled a finger, patterns got recorded on an EEG amplifier. Green had had to work out a means of separating out the true effects of healing from this electrostatic noise. The only way to do so, as he saw it, was to have his healers remain perfectly still while they were sending out healing energy.

Schwartz listened to the talk with growing fascination. Green was discarding what might be the most interesting part of the data, he thought. *One man's noise was another man's signal.* Does movement, even the physiology of your breathing, create an electromagnetic signal big enough to be picked up on a copper wall? Could it be that human beings were not only receivers of signals but also *transmitters*?

It made perfect sense that we transmitted energy. A great deal of evidence had already proved that all living tissue has an electric charge. Placing this charge in three-dimensional space caused an electromagnetic field that travelled at the speed of light. The mechanisms for the transmission of energy were clear, but what was unclear was the degree to which we sent out electromagnetic fields just by simple movements and whether our energy was being picked up by other living things.

Schwartz was itching to test this out for himself. After the conference, he contacted Green for advice about how to build his own copper wall. He rushed to Home Depot, which did not stock copper shielding but did have aluminium shielding, which could also act as a rudimentary antenna. He purchased some two by fours, placed them on glass bricks so that they would be isolated from the ground, and used them to assemble a 'wall'. After he had attached the wall to an EEG amplifier, he began playing around with the effects of his hand, waving it back and forth above the box. As he suspected, the amplifier tracked the movement. His hand movements were generating signals.[4]

Schwartz began demonstrating these effects in front of his students in his faculty office, making use of a bust of Einstein for dramatic effect. With these experiments, he made use of an EEG cap, with its dozens of electrodes. When not picking up brain signals, the cap will register only noise on the amplifier.

During his experiments, Schwartz placed the EEG cap on his Einstein bust, and turned on just a single electrode channel on the top of the cap. Then he moved his hand over Einstein's head. As though the great man had suddenly experienced a moment of enlightenment, the amplifier suddenly came alive and produced evidence of an electromagnetic wave. But the signal, Schwartz explained to his students, was not a sudden brain wave emitted from the lifeless statue – only the tracking of the electromagnetic field produced by his arm's movement. It seemed indisputable: his body must be sending out a signal with every single flutter of his hand.

Schwartz got more creative with his experiments. When he tried the same gesture from three feet away, the signal diminished. When he placed the bust in a Faraday cage, an enclosure of tightly knit copper mesh that screens out electromagnetic fields, all effect disappeared. This strange energy resulting from movement had all the hallmarks of electricity: it decreased with distance, and was blocked by an electromagnetic shield.

At one point, Schwartz asked one of the students to stand with his left hand over Einstein's head, with his right arm extended towards Schwartz, who was sitting in a chair three feet away. Schwartz moved his arm up and down. To the amazement of the other students, Schwartz's movement was picked up by the amplifier. The signal had passed through Schwartz's body and travelled through the student. Schwartz was still generating the signal, but this time, the student had become the antenna, receiving the signal and transmitting it to the amplifier, which acted as another antenna.

Schwartz realized he had hit upon the most important point of all his research. Simple movement generated electrical charge, but, more important, it created a relationship. Every movement we make appears to be felt by the people around us. The implications were staggering. What if he were admonishing a student? What might be the physical effect on the student of wagging his finger while shouting 'Don't do that'? The student might feel as if he were getting shot with a wave of energy. Some people might even have more powerful positive or negative charges than others. In Elmer Green's copper wall experiment, all sorts of equipment malfunctioned in the presence of Roslyn Bruyere, a famous healer.

Schwartz was onto something fundamental about the actual energy that human beings emit. Could the energy of thought have the same effect as the energy of movement outside the thinker's own body? Did thoughts also create a relationship with the people around us? Every intention towards someone else might have its own physical counterpart, which would be registered by its recipient as a physical effect.

Like Schwartz, I suspected the energy generated by thoughts did not behave in the same way as the energy generated by movement.

After all, the signal from movement decreased over distance, much like ordinary electricity. With healing, distance appeared to be irrelevant. The energy of intention, if indeed there were any, would have to be more fundamental than that of ordinary electromagnetism – and lie somewhere, perhaps, in the realm of quantum physics. How could I test the energetic effects of intention? Healers, who appeared to be sending more energy than normal through their healing, offered an obvious place to start.

Elmer Green demonstrated in his research that an enormous surge of electrostatic energy occurred during healing. When a person is simply standing still, his or her breathing and beating heart will produce electrostatic energy of 10–15 millivolts on the EEG amplifiers; during activities requiring focused attention, such as meditation, the energy will surge up to 3 volts. During healing, however, Green's healers produced voltage surges up to 190 volts; one produced 15 such pulses, which were 100,000 times higher than normal, with smaller pulses of 1–5 volts appearing on each of the four copper walls. On investigating the source of this energy, Green discovered that the pulses were coming from the healer's abdomen, called *dan tien* and considered the central engine of internal energy in the body in Chinese martial arts.[5]

Stanford University physicist William Tiller constructed an ingenious device to measure the energy produced by healers. The equipment discharged a steady stream of gas and recorded the exact number of electrons pulsing out with the discharge. Any increase in voltage would be captured by the pulse counter.

In his experiment, Tiller asked ordinary volunteers to place their hands about six inches from his device and hold a mental intention to increase the count rate. In the majority of more than 1000 such experiments, Tiller discovered that, during the intention, the number of recorded pulses would increase by 50,000 and remain there for 5 minutes. These increases would occur even if a participant was not close to the machine, so long as he or she held an intention.

Tiller concluded that directed thoughts produce demonstrable physical energy, even over remote distance.[6]

I found two other studies measuring the actual electrical frequencies emitted by people using intention. One study measured healing energy and the other examined energy generated by a Chinese *Qigong* master during times that he was emitting external *Qi*, the Chinese term for energy or the life force.[7] In both instances, the measurements were identical: frequency levels of 2–30 hertz were being emitted by the healers.

This energy also seemed to change the molecular nature of matter. I discovered a body of scientific evidence examining chemical changes caused by intention. Bernard Grad, an associate professor of biology at McGill University in Montreal, had examined the effect of healing energy on water that was to be used to irrigate plants. After a group of healers had sent healing to samples of water, Grad chemically analysed the water by infrared spectroscopy. He discovered that the water treated by the healers had undergone a fundamental change in the bonding of oxygen and hydrogen in its molecular makeup. The hydrogen bonding between the molecules had lessened in a similar manner to that which occurs in water exposed to magnets.[8] A number of other scientists confirmed Grad's findings; Russian research discovered that the hydrogen–oxygen bonds in water molecules undergo distortions in the crystalline microstructure during healing.[9]

These kinds of changes can occur simply through the act of intention. In one study, experienced meditators sent an intention to affect the molecular structure of water samples they were holding throughout the meditation. When the water was later examined by infrared spectrophotometry, many of its essential qualities, particularly its absorbance – the amount of light absorbed by the water at a particular wavelength – had been significantly altered.[10] When someone holds a focused thought, he may be altering the very molecular structure of the object of his intention.

In his research, Gary Schwartz wondered whether intention only manifested as electrostatic energy. Perhaps magnetic energy also played a role. Magnetic fields naturally had more power, more 'push–pull' energy. Magnetism seemed the more powerful and universal energy; the earth itself is profoundly influenced by its own faint pulse of geomagnetic energy. Schwartz remembered a study carried out by William Tiller, in which psychics had been placed inside a variety of devices that block different forms of energy. They had performed better than usual in a Faraday cage, which filters out only electrical energy, but they performed worse when placed in a magnetically shielded room.[11]

From these early studies, Schwartz gleaned two important implications: healing may generate an initial surge of electricity, but the real transfer mechanism may be magnetic. Indeed, psychic phenomena and psychokinesis could be differentially influenced, simply through different types of shielding. Electrical signals might interfere, while magnetic signals enhance the process.

To test this latest idea, Schwartz was approached by a colleague of his, Melinda Connor, a post-doctoral fellow in her mid-forties with an interest in healing. The first hurdle was finding an accurate means of picking up magnetic signals. Measuring tiny low-frequency magnetic fields is tricky, requiring the use of expensive and highly sensitive equipment called a SQUID, or superconducting quantum interference device. A SQUID, which can cost up to four million dollars, ordinarily occupies a specially constructed room that has been magnetically shielded in order to eliminate ambient radiating noise.

The best Schwartz and Connor could come up with on their limited budget was a poor man's SQUID – a small handheld, battery-operated three-axis digital gaussmeter originally designed to measure electromagnetic pollution by picking up extra-low-frequency (ELF) magnetic fields. The gaussmeter was sensitive enough to pick up one-thousandth of a gauss, a very faint pulse of a magnetic field. In

Schwartz's mind, this level of sensitivity was more than adequate to do the job.

It occurred to Connor that the way to measure change in low-frequency magnetic fields was to count the number of changes in the meter reading over time. When simply recording ambient stable magnetic fields, the device will only deviate slightly – by less than one-tenth of a gauss. However, in the presence of an oscillating magnetic field – with periodic changes in frequency – the numbers will keep moving, from, say, 0.6 to 0.7 to 0.8, and back down to 0.6. The greater and more frequent the change, which would be recorded by the number of changes in the dials, the more likely it is that the magnetic field has been affected by a source of directed energy.

Connor and Schwartz gathered together a group of practitioners of Reiki, the healing art developed a century ago in Japan. They took measurements near each hand of all the healers during alternating periods while they were 'running energy' and then during times they were at rest, with their eyes closed. Next, the pair assembled a group of 'master healers' with a substantial track record of successful, dramatic healings. Again, Connor and Schwartz took magnetic field measurements near each hand, while the master healers were running energy and at rest. Then, they compared the Reiki measurements with measurements they had taken of people who had not been trained in healing.

Once Schwartz and Conner had analysed the data, they discovered that both groups of healers demonstrated significant fluctuations in very low pulsations of a magnetic field, emanating from both hands. A huge increase in oscillations in the magnetic field occurred whenever a healer began to run energy. However, the most profound energy increase surged from their dominant hands. The control group of people who were not trained healers did not demonstrate the same effect.

Then Schwartz compared effects from the Reiki group with those of the master healers and discovered another enormous difference.

The master healers averaged close to a third more magnetic-field changes per minute than the Reiki healers.[12]

The study results seemed clear. Schwartz and Connor had their proof that directed intention manifests as both electrostatic *and* magnetic energy. But they also discovered that intention was like playing the piano; you need to learn how to do it, and some people do it better than others.

In considering what this all meant, Gary Schwartz thought of the phrase often used by medical doctors, usually in emergency situations: *when you hear hoof beats, don't think zebras*. In other words, when you are trying to diagnose someone with physical symptoms, first rule out all the most likely causes, and only then consider more exotic possibilities. He liked to approach science in the same way and so he questioned his own findings: Could the healers' increase in magnetic field oscillations during healing simply be the result of certain peripheral biophysical changes? Muscle contractions generate a magnetic field, as do changes in blood flow, the increasing or decreasing dilation of blood vessels, the body's current volume of liquid or even the flow of electrolytes. Skin, sweat glands, change of temperature, neural induction – all generate magnetic fields. His guess was that healing resulted from a summation of multiple biological processes that are mediated magnetically.

But the possibility that healing might be a magnetic effect did not explain long-distance remote healing. In some instances, healers sent healing from thousands of miles away and the effect did not decay with distance. In one successful study of AIDS patients who improved through remote healing, the 40 healers involved in the study sent the healing to the San Francisco patients from locations all across America.[13] Similar to electrical fields, magnetic fields decrease with distance. The magnetic and electrical effects were likely to be some aspect of the process, but not its central one. It was likely to be closer to a quantum field, possibly more akin to light.

Schwartz began to consider the possibility that the mechanism creating intention originated with the tiny elements of light emitted from human beings. In the mid-1970s, a German physicist named Fritz-Albert Popp had stumbled upon the fact that all living things, from the most basic of single-celled plants to the most sophisticated of organisms like human beings, emitted a constant tiny current of photons – tiny particles of light.[14] He labelled them 'biophoton emissions' and believed that he had uncovered the primary communication channel of a living organism – that it used light as a means of signalling to itself and to the outside world.

For more than 30 years, Popp has maintained that this faint radiation, rather than biochemistry, is the true driving force in orchestrating and coordinating all cellular processes in the body. Light waves offered a perfect communication system able to transfer information almost instantaneously across an organism. Having waves, rather than chemicals, as the communication mechanism of a living being also solved the central problem of genetics – how we grow and take final shape from a single cell. It also explains how our bodies manage to carry out tasks with different body parts simultaneously. Popp theorized that this light must be like a master tuning fork setting off certain frequencies that would be followed by other molecules of the body.[15]

A number of biologists, such as the German biophysicist Herbert Fröhlich, had proposed that a type of collective vibration causes proteins and cells to coordinate their activities. Nevertheless, all such theories were ignored until Popp's discoveries, largely because no equipment was sensitive enough to prove they were right.

With the help of one of his students, Popp constructed the first such machine – a photomultiplier that captured light and counted it, photon by photon. He carried out years of impeccable experimentation that demonstrated that these tiny frequencies were mainly stored and emitted from the DNA of cells. The intensity of the light in organisms was stable, ranging from a few to several hundred photons per second per square centimetre surface of the living thing – until the organism was somehow disturbed or ill, at which point the current went sharply up or down. The signals contained valuable information about the state of the body's health and the effects of any particular therapy. Cancer victims had fewer photons, for instance. It was almost as though their light were going out.

Initially vilified for his theory, Popp was eventually recognized by the German government and then internationally. Eventually he formed the International Institute of Biophysics (IIB), composed of 15 groups of scientists from international centres all around the world, including prestigious institutions like CERN in Switzerland, Northeastern University in the USA, the Institute of Biophysics Academy of Science in Beijing, China, and Moscow State University in Russia. By the early twenty-first century, the IIB numbered at least 40 distinguished scientists from around the globe.

Could it be that these were the frequencies that mediated healing? Schwartz realized that if he was going to carry out studies of biophoton emissions, first he had to figure out how to view these tiny emissions of light. In his laboratory, Popp developed a computerized mechanism attached to a box in which a living thing, such as a plant, could be placed. The machine could count the photons and chart the amount of light emitted on a graph. But those machines only recorded photons in utter pitch blackness. Up until then, it had been impossible for scientists to witness living things actually glowing in the dark.

As Schwartz mulled over the kind of equipment that would allow him to see very faint light, he thought of state-of-the-art supercooled charge-coupled device (CCD) cameras on telescopes. This exquisitely sensitive equipment, now used to photograph galaxies deep in space, picks up about 70 per cent of any light, no matter how faint. CCD devices were also used for night-vision equipment. If a CCD camera could pick up the light from the most distant stars, it might also be able to pick up the faint light coming off living things.

However, this kind of equipment can cost hundreds of thousands of dollars and usually had to be cooled to temperatures only 100 degrees above absolute zero, to eliminate any ambient radiation emitted at room temperature. Cooling the camera down also helped to improve its sensitivity to faint light. Where on earth was he going to get hold of this kind of high-tech equipment?

Kathy Creath, a professor of optical sciences at Schwartz's university, who shared his fascination with living light and its possible role in healing, had an idea. As it happened, she knew that the department of radiology at the National Science Foundation (NSF) in Tucson owned a low-light CCD camera, which they used to measure the light emitted from laboratory rats after being injected with phosphorescent dyes. The Roper Scientific VersArray 1300 B low-noise, high-performance CCD camera was housed in a dark room inside a black box and above a Cryotiger cooling system, which cooled temperatures to $-100°C$. A computer screen displayed its images. It was just what they were looking for. After Creath approached the director of the NSF project, he generously agreed to allow the two of them access to the camera during its down time.

In their first test, Schwartz and Creath placed a geranium leaf on a black platform. They took fluorescent photographs after exposures of up to five hours. When the computer displayed the final photograph, it was dazzling: a perfect image of the leaf in light, like a shadow in reverse, but in incredible detail, each of its tiniest veins delineated. Surrounding the leaf were little white spots, like a sprinkling of fairy dust – evidence of high-energy cosmic rays. With his next exposure, Schwartz used a software filter to screen out the ambient radiation. The image of the leaf was now perfect.

As they studied this latest photograph on the screen of the computer in front of them, Schwartz and Creath understood that they were making history. It was the first time a scientist had been able to witness images of the light actually emanating from a living thing.[16]

Now that he had equipment that captured and recorded light, Schwartz was finally able to test whether healing intention also generated light. Creath got hold of a number of healers, and asked them to place their hands on the platform underneath the camera for 10 minutes. Schwartz's first crude images showed a rough glow of large pixilations, but they were too out of focus for him to analyse them. Next he tried placing the healers' hands on a white background (which reflected light) rather than on a black background (which absorbed light). The images were breathtakingly clear: a stream of light flowed out of the healers' dominant hands, almost as though it were flowing from their fingers. Schwartz now had his answer about the nature of conscious thought: healing intention creates waves of light – and, indeed, among the most organized light waves found in nature.

The theory of relativity was not Einstein's only great insight. He had had another astonishing realization in 1924, after correspondence with an obscure Indian physicist, Satyendra Nath Bose, who had been pondering the then-new idea that light was composed of little vibrating packets called photons. Bose had worked out that, at certain points, photons should be treated as identical particles. At the time nobody believed him – nobody but Einstein, after Bose sent him his calculations.

Einstein liked Bose's proofs and used his influence to get Bose's theory published. Einstein also was inspired to explore whether, under certain conditions or certain temperatures, atoms in a gas, which ordinarily vibrated anarchically, might also begin to behave in synchrony, like Bose's photons. Einstein set to work on his own formula to determine which conditions might create such a phenomenon. When he reviewed his figures, he thought he had made a mistake in his calculations. According to his results, at certain extraordinarily low temperatures, just a few kelvin above absolute zero, something really strange would begin to happen: the atoms, which ordinarily can operate at a number of different speeds, would slow down to identical energy levels. In this state, the atoms would lose their individuality and both look and behave like one giant atom. Nothing in his mathematical armamentarium could tell them apart. If this were true, he realized, he had stumbled upon an entirely new state of matter, with utterly different properties from anything known in the universe.

Einstein published his findings,[17] and lent his name to the phenomenon, called a Bose–Einstein condensate, but he was never convinced that he had been right. Nor were other physicists, until more than 70 years later when, on 5 June 1995, Eric Cornell and Carl Wieman of JILA, a programme sponsored by the National Institute of Standards and Technology and the University of Colorado at Boulder, managed to cool a tiny batch of rubidium atoms down to 170 billionths of a degree above absolute zero.[18] It had been quite a feat, requiring trapping the atoms in a web of laser light and then magnetic fields. At a certain point, a batch of some 2000 atoms – measuring about 20 microns, about one-fifth the thickness of a single piece of paper – began behaving differently from the cloud of atoms surrounding them, like one smeared-out single entity. Although the atoms were still part of a gas, they were behaving more like the atoms of a solid.

Four months later, Wolfgang Ketterle from Massachusetts Institute of Technology replicated their experiment, but with a form of sodium, for which he, as well as Cornell and Wieman, won the 2001 Nobel prize.[19] Then a few years after that, Ketterle and others like him were able to reproduce the effect with molecules.[20]

Scientists believed that a form of Einstein and Bose's theory could account for some of the strange properties they had begun to observe in the subatomic world: superfluidity, when certain fluids can flow without losing energy, or even spontaneously work themselves out of their containers; or superconduction, a similar property of electrons in a circuit. In superfluid or superconductor states, liquid or electricity could theoretically flow at the same pace forever.

Ketterle had discovered another amazing property of atoms or molecules in this state. All the atoms were oscillating in perfect harmony, similar to photons in a laser, which behave like one giant photon, vibrating in perfect rhythm. This organization makes for an extraordinary efficiency of energy. Instead of sending a light about 3 metres, the laser emits a wave 300 million times that far.

Scientists were convinced that a Bose–Einstein condensate was a peculiar property of atoms and molecules slowing down so much that they are almost at rest, when exposed to temperatures only a fraction above the coldest temperatures in the universe. But then Fritz–Albert Popp and the scientists working with him made the astonishing discovery that a similar property existed in the weak light emanating from organisms. This was not supposed to happen in the boiling inner world of the living thing. What is more, the biophotons he measured from plants, animals and humans were highly coherent. They acted like a single super-powerful frequency, a phenomenon also referred to as 'superradiance'.

The German biophysicist Herbert Fröhlich had first described a model in which this type of order could be present and play a central role in biological systems. His model showed that, with complex dynamic systems like human beings, the energy within created all sorts of subtle relationships, so that it is no longer discordant.[21] Living energy is able to organize to one giant coherent state, with the highest form of quantum order known to nature. When subatomic particles are said to be 'coherent', or 'ordered', they become highly interlinked by bands of common electromagnetic fields, and resonate like a multitude of tuning forks all attuned to the same frequency. They stop behaving like anarchic individuals and begin operating like one well-rehearsed marching band.

As one scientist put it, coherence is like comparing the photons of a single 60-watt light bulb to the sun. Ordinarily, light is extraordinarily inefficient. The intensity of light from a bulb is only about 1 watt per square centimetre of light – because many of the waves made by the photons destructively interfere with and cancel out each other. The light per square centimetre generated by the sun is about 6000 times stronger. But if you could get all the photons of this one small light bulb to become coherent and resonate in harmony with each other, the energy density of the single light bulb would be thousands to millions of times higher than that of the surface of the sun.[22]

After Popp made his discoveries about coherent light in living organisms, other scientists postulated that mental processes also create Bose–Einstein condensates. British physicist Roger Penrose and his partner, American anaesthetist Stuart Hameroff from the University of Arizona, were in the vanguard of frontier scientists who proposed that the microtubules in cells, which create the basic structure of the cells, were 'light pipes' through which disordered wave signals were transformed into highly coherent photons and pulsed through the rest of the body.[23]

Gary Schwartz had witnessed just this coherent photon stream emanating from the hands of healers. After studying the work of scientists like Popp and Hameroff, he finally had his answer about the source of healing: if thoughts are generated as frequencies, healing intention is well-ordered light.

Gary Schwartz's creative experiments revealed to me something fundamental about the quantum nature of thoughts and intentions. He and his colleagues had uncovered evidence that human beings are both receivers and transmitters of quantum signals. Directed intention appears to manifest as both electrical and magnetic energy and to produce an ordered stream of photons, visible and measurable by sensitive equipment. Perhaps our intentions also operate as highly coherent frequencies, changing the very molecular makeup and bonding of matter. Like any other form of coherence in the subatomic world, one well-directed thought might be like a laser light, illuminating without ever losing its power.

I was reminded of an extraordinary experience Schwartz once had in Vancouver. He had been staying in the penthouse apartment suite of a downtown hotel. He had awakened at 2 a.m., as he often did, and had walked out to the balcony to have a look at the spectacular view of the city to the west, framed by the mountains. He was surprised to see how many hundreds of homes along the peninsula below him still had their lights on. He wished he had a telescope handy to see what some of the people were doing up at this late hour. But of course, if any of them had their own telescope, they would be able to see him standing there in the nude. An odd thought suddenly came to him of his own naked image flying into each window. But maybe the idea was not so fanciful. After all, he was emitting a constant stream of biophotons, all travelling at the speed of light; each photon would have travelled 186,000 miles one second later, and 372,000 miles one second after that.

His light was not unlike the photons of visible light emanating from stars in the sky. Much of the light from distant stars has been travelling for millions of years. Starlight contains a star's individual history. Even if a star had died long before its light reached earth, its information remains, an indelible footprint in the sky.

He then had a sudden image of himself as a ball of energy fields, a little star, glowing with a steady stream of every photon his body had ever produced for more than 50 years. All the information he had been sending from the time he was a young boy in Long Island, every last thought he had ever had, was still out there, glowing like starlight. Perhaps, I thought, intention was also like a star. Once constructed, a thought radiated out like starlight, affecting everything in its path.

CHAPTER THREE

The Two-Way Street

CLEVE BACKSTER WAS AMONG THE FIRST to propose that plants are affected by human intention – a notion considered so preposterous that it was ridiculed for 40 years. Backster achieved his notoriety from a series of experiments that purported to demonstrate that living organisms read and respond to a person's thoughts.

Plant telepathy interested me less than a tangential discovery of his that has been sidelined amid all his adverse publicity: evidence of a constant two-way flow of information between all living things. Every organism, from bacteria to human beings, appears to be in perpetual quantum communication. This relentless conversation offers a ready mechanism by which thoughts can have a physical effect.

This discovery resulted from a silly little diversion in 1966; Backster, at the time a tall, wiry man with a buzz cut and a great deal of childlike enthusiasm, was easily distracted. He often carried on working in his suite of offices when the rest of his staff had gone home and he could finally focus without the constant interruptions of colleagues and the tumultuous daytime activity of Times Square, four storeys below.[1]

Backster had made his name as the country's leading lie-detector expert. During the Second World War, he had been fascinated by the psychology of lying, and the use of hypnosis and 'truth serum' inter-rogation in counter-intelligence, and he had brought these twin fascin-ations to bear in refining the polygraph test to a high psychological art. He had launched his first programme with the CIA for counter-intelligence several years after the war, and then went on to found the Backster School of Lie Detection, still the world's leading school teaching polygraph techniques some 50 years after it first opened its doors.

One morning in February, after working all night, Backster was taking a coffee break at 7 a.m. He was about to water the *Dracaena* and rubber plant in his office. As he filled up his watering can, he wondered if it might be possible to measure the length of time it would take water to travel up the stem of a plant from the roots and reach the leaves, particularly in the *Dracaena*, a cane plant with an especially long trunk. It occurred to him that he could test this by connecting the *Dracaena* to one of his polygraph machines; once the water reached the spot between the electrodes, the moisture would contaminate the circuit and be recorded as a drop in resistance.

A lie detector is sensitive to the slightest change in the electrical conductivity of skin, which is caused by increased activity of the sweat glands, which in turn are governed by the sympathetic nerv-ous system. The polygraph galvanic skin response (GSR) portion of the test displays the amount of the skin's electrical resistance, much as an electrician's ohmmeter records the electrical resistance of a circuit. A lie detector also monitors changes in blood pressure, respiration, and the strength and rate of the pulse. Low levels of electrical conductivity indicate little stress and a state of calm. Higher electrodermal activity (EDA) readings indicate that the sympathetic nervous system, which is sensitive to stress or certain emotional states, is in overdrive – as would be the case when some-one is lying. A polygraph reading can offer evidence of stress to the sympathetic nervous system even before the person being tested is consciously aware of it.

In 1966, the state-of-the-art technology consisted of a set of electrode plates, which were attached to two of a subject's fingers, and through which a tiny current of electricity was passed. The smallest increases or decreases in electrical resistance were picked up by the plates and recorded on a paper chart, on which a pen traced a continuous, serrated line. When someone lied or in any way experienced a surge of emotion (such as excitement or fear), the size of the zigzag would dramatically increase and the tracing would move to the top of the chart.

Backster sandwiched one of the long, curved leaves of the *Dracaena* between the two sensor electrodes of a lie detector and encircled it with a rubber band. Once he watered the plant, what he expected to see was an upward trend in the ink tracing on the polygraph recording paper, corresponding to a drop in the leaf's electrical resistance as the moisture content increased. But as he poured in the water, the very opposite occurred. The first part of the tracing began heading downward and then displayed a short-term blip, similar to what happens when a person briefly experiences a fear of detection.

At the time Backster thought he was witnessing a human-style reaction, although he would later learn that the waxy insulation between the cells in plants causes an electrical discharge that mimics a human stress reaction on polygraph instruments. He decided that if the plant were indeed displaying an emotional reaction, he would have to come up with some major emotional stimulus to heighten this response.

When a person takes a polygraph test, the best way to determine if he is lying is to ask a direct and pointed question, so that any answer but the truth will cause an immediate, dramatic stress reaction in his sympathetic nervous system: 'Was it you who fired the two bullets into Joe Smith?'

In order to elicit the equivalent of alarm in a plant, Backster knew he needed somehow to threaten its well-being. He tried immersing one of the plant's leaves in a cup of coffee, but that did not cause any interesting reaction on the tracing – only a continuation of the downward trend. If this were the tracing of a human being, Backster would have concluded that the person being monitored was tired or bored. It was obvious to him that he needed to pose an immediate and genuine threat: he would get a match and burn the electroded leaf.

At the very moment he had that thought, the recording pen swung to the top of the polygraph chart and nearly jumped off. He had not burned the plant; he had only *thought* about doing so. According to his polygraph, the plant had perceived the thought as a direct threat and registered extreme alarm. He ran to his secretary's desk in a neighbouring office for some matches. When he returned, the plant was still registering alarm on the polygraph. He lit a match and flickered it under one of the leaves. The pen continued on its wild, zigzag course. Backster then returned the matches to his secretary's desk. The tracing calmed down and began to flat-line.

He hadn't known what to make of it. He had long been drawn to hypnosis and ideas about the power of thought and the nature of consciousness. He had even performed a number of experiments with hypnosis during his work with the Army Counter Intelligence Corps and the CIA, as part of a campaign designed to detect the use of hypnosis techniques in Russian espionage.

But this was something altogether more extraordinary. This plant, it seemed, had *read his thoughts*. It wasn't even as though he particularly *liked* plants. This only could have occurred if the plant possessed some sort of sophisticated extrasensory perception. The plant somehow must be attuned to its environment, able to receive far more than pure sensory information from water or light.

Backster modified his polygraph equipment to amplify electrical signals so that they would be highly sensitive to the slightest electrical change in the plants. He and his partner, Bob Henson, set about replicating the initial experiment. Backster spent the next year and a half frequently monitoring the reactions of the other plants in the office to their environment. They discovered a number of characteristics. The plants grew attuned to the comings and goings of their main caretaker. They also maintained some sort of 'territoriality' and so did not react to events in the other offices near Backster's lab. They even seemed to tune in to Pete, his Doberman Pinscher, who spent his days at the office.

Most intriguing of all, there seemed to be a continuous two-way flow of information between the plants and other living things in their environment. One day, when Backster boiled his kettle to make coffee, he found he had put in too much water. But when he poured the residue down the sink, he noticed that the plants registered an intense reaction.

The sink was not the most hygienic; indeed, his staff had not cleaned the drain for several months. He decided to take some samples from the drain and examine them under a microscope, which showed a jungle of bacteria that ordinarily lives in the waste pipes of a sink. When threatened by the boiling water, had the bacteria emitted a type of mayday signal before they died, which had been picked up by the plants?

Backster, who knew he would be ridiculed if he presented findings like these to the scientific community, enlisted an impressive array of chemists, biologists, psychiatrists, psychologists and physicists to help him design an airtight experiment. In his early experiments, Backster had relied upon human thought and emotion as the trigger for reactions in the plants. The scientists discouraged him from using intention as the stimulus of the experiment, because it did not lend itself to rigorous scientific design. How could you set up a control for a human thought – an intention to harm, say? The orthodox scientific community could easily pick holes in his study. He had to create a laboratory barren of any other living things besides the plants to ensure that the plants would not be, as it were, distracted.

The only way to achieve this was to automate the experiment entirely. But he also needed a potent stimulus. He tried to think of the one act that would stir up the most profound reaction, something that would evoke the equivalent in the plants of dumbfounded horror. It became clear that the only way to get unequivocal results was to commit the equivalent of mass genocide. But what could he kill *en masse* that would not arouse the ire of anti-vivisectionists or get him arrested? It obviously could not be a person or a large animal of any variety. He did not even want to kill members of the usual experimental population, like rats or guinea pigs. The one obvious candidate was brine shrimp. Their only purpose, as far as he could tell, was to become fodder for tropical fish. Brine shrimp were already destined for the slaughterhouse. Only the most ardent anti-vivisectionist could object.

Backster and Henson rigged up a gadget that would randomly select one of six possible moments when a small cup containing the brine shrimp would invert and tip its contents into a pot of continuously boiling water. The randomizer was placed in the far room in his suite of six offices, with three plants attached to polygraph equipment in three separate rooms at the other end of the laboratory. His fourth polygraph machine, attached to a fixed valve resistor to ensure that there was no sudden surge of voltage from the equipment, acted as the control.

Microcomputers had yet to be invented, as Backster set up his lab in the late sixties. To perform the task, Backster created an innovative mechanical programmer, which operated on a time-delay switch, to set off each event in the automation process. After flipping the switch, Backster and Henson would leave the lab, so they and their thoughts would not influence the results. He had to eliminate the possibility that the plants might be more attuned to him and his colleague than a minor murder of brine shrimp down the hallway.

Backster and Henson tried their test numerous times. The results were unambiguous: the polygraphs of the electroded plants spiked a significant number of times just at the point when the brine shrimp hit the boiling water. Years after he had made this discovery – and after he became a great fan of *Star Wars* – he would think of this moment as one in which his plants picked up a major disturbance in the Force, and he had discovered a means of measuring it.[2] If plants could register the death of an organism three doors away, it must mean that all life forms were exquisitely in tune with each other. Living things must be registering and passing telepathic information back and forth at every moment, particularly at moments of threat or death.

Backster published the results of his experiment in several respected journals of psychic research and gave a modest presentation before the Parapsychology Association during its tenth annual meeting.[3] Parapsychologists recognized Backster's contribution and replicated it in a number of independent laboratories, notably that of Alexander Dubrov, a Russian doctor of botany and plant physiology.[4] It was even glorified in a bestselling book, *The Secret Life of Plants*.[5] But among the mainstream scientific community, his research was disparaged as ludicrous, largely because he was not a traditional scientist, and he was ridiculed for what became known as 'The Backster Effect'. In 1975, *Esquire* magazine even awarded him one of its 100 Dubious Achievement Awards: 'Scientist claims yogurt talks to itself'.[6]

Nonetheless, over the next 30 years Backster ignored his critics and stubbornly carried on with his research, as well as his polygraph business, eventually amassing file drawers full of studies of what he referred to as 'primary perception'. A variety of plants that had been hooked up to his polygraph equipment showed evidence of a reaction to human emotional highs and lows, especially threats and other forms of negative intention – as did paramecia, mould cultures, eggs and, indeed, yogurt.[7] Backster even demonstrated that bodily fluids such as blood and semen samples taken from himself and his colleagues registered reactions mirroring the emotional state of their hosts; the blood cells of a young lab assistant reacted intensely the moment he opened a *Playboy* centrefold and caught sight of Bo Derek in the nude.[8]

These reactions were not dependent on distance; any living system attached to a polygraph reacted similarly to his thoughts, whether he was in the room or miles away. Like pets, they had become attuned to their 'owner'. These organisms were not simply registering his thoughts; they were communicating telepathically with all the living things in their environment. The live bacteria in yogurt displayed a reaction to the death of other types of bacteria and even evidenced a desire to be 'fed' with more of its own beneficial bacteria. Eggs registered a cry of alarm and then resignation when one of their number was dropped in boiling water. Plants appeared to react in real time to any break in continuity with the living beings in their environment. They even appeared to react at the moment when their caretakers, who were away from the office, decided to return.[9]

His major difficulty was designing experiments that could demonstrate an effect scientifically. Even though his laboratory experiments were now entirely automated, when he left the office, the plants would remain attuned to him, no matter now far away he went. If Backster and his partner were at a bar a block away during an experiment, he would discover that the plants were not responding to the brine shrimp, but to the rising and falling animation of their conversations. It got so difficult to isolate reactions to specific events that eventually he had to design experiments that would be carried out by strangers in another lab.

Repeatability remained another big problem. Any tests required spontaneity and true intent. He had discovered this when the famous remote viewer Ingo Swann had come to visit him at his lab in October 1971. Swann wanted to repeat Backster's initial experiment with his *Dracaena*. As expected, the plant's polygraph began to spike when Swann imagined burning the plant with a match. He tried it again, and the plant reacted wildly, then stopped.

'What does *that* mean?' Swann asked.

Backster shrugged. 'You tell me.'

The thought that occurred to Swann was so bizarre that he was not sure whether to say it aloud. 'Do you mean,' he said, 'that it has *learned* that I'm not serious about really burning its leaf? So that it now knows it need not be alarmed?'

'*You* said it, I didn't,' Backster replied. 'Try another kind of harmful thought.'

Swann thought of putting acid in the plant's pot. The needle on the polygraph again began to zigzag wildly. Eventually, the plant appeared to understand that Swann was not serious. The polygraph tracing flat-lined. Swann, a plant lover who was already convinced that plants were sentient, was nevertheless shocked at the thought that plants could learn to differentiate between true and artificial human intent: *a plant learning curve.*[10]

Although certain questions remain about Backster's unorthodox research methods, the sheer bulk of his evidence argues strongly for some sort of primary responsiveness and attuning, if not sentience, present in all organisms, no matter how primitive. But for my purposes, Backster's real contribution was his discovery of the telepathic communication carrying on between every living thing and its environment. Somehow, a constant stream of messages was being sent out, received and replied to.

Backster had to wait some years to discover the mechanism of this communication, which became apparent when physicist Fritz-Albert Popp discovered biophotons.[11] At first Popp believed that a living organism used biophoton emissions solely as a means of instantaneous, non-local signalling from one part of the body to another – to send information about the global state of the body's health, say, or the effects of any particular treatment. But then Popp grew intrigued by the most fascinating effect of all: the light seemed to be a communications system *between* living things.[12] In experiments with *Daphnia*, a common water flea, he discovered that female water fleas were absorbing the light emitted from each other and sending back wave interference patterns, as though they had taken the light sent to themselves and updated it with more information. Popp concluded that this activity may be the mechanism enabling fleas to stay together when they swarm – a silent communication holding them together like an invisible net.[13]

He decided to examine the light emissions between dinoflagellates, luminescent algae that cause phosphorescence in seawater. These single-celled organisms sit somewhere between an animal and a plant in the evolutionary scale; although they are classified as a plant, they move like a primitive animal. Popp discovered that the light of each dinoflagellate was coordinated with that of its neighbours, as if each were holding aloft a tiny lantern on cue.[14] Chinese colleagues of Popp's who had tried positioning two samples of the algae so that they could 'see' each other through a shutter also found that the light emissions from each sample were synchronous. The researchers concluded that they had witnessed a highly sophisticated means of communication. There was no doubt that the two samples were signalling to each other.[15]

These organisms also appeared to be registering light from other species, although the greatest synchronicities occurred between members of the same species.[16] Once the light waves of one organism were initially absorbed by another organism, the first organism's light would begin trading information in synchrony.[17] Living things also appeared to communicate information with their surroundings. Bacteria absorbed light from their nutritional media: the more bacteria present, Popp found, the greater the absorption of light.[18] Even the white and yolk of an egg appear to communicate with the shell.[19]

This communication carries on, even if an organism is cut into pieces. Gary Schwartz cut up a batch of string beans, placed them between 1 millimetre and 10 millimetres apart, and then used the NSF CCD camera he had borrowed to take a series of photographs of the sections. Using software to enhance the light between the beans, he discovered so much light between the sections that it appeared as though the bean were whole again. Even though the string beans had been severed, the individual sections carried on their communication to the rest of the vegetable.[20] This may be the mechanism accounting for the feeling described by amputees with phantom limb sensations. The light of the body still communicates with the energetic 'footprint' of the amputated limb.

Like Backster, Popp discovered that living things are exquisitely in tune with their environment through these light emissions. One of Popp's colleagues, Professor Wolfgang Klimek, the head of the Ministry of Research for the German government, devised an ingenious experiment to examine whether creatures such as algae were aware of past disturbances in their environment. He prepared two containers of seawater, and shook one of them. After 10 minutes, when the water in the shaken container had settled down, he placed samples of dinoflagellates in the two vessels. Those algae exposed to the shaken water suddenly increased their photon emissions – a sign of stress. The algae appeared to be aware of the slightest change in their environment – even a historical change – and responded with alarm.[21]

Another of Popp's colleagues, Eduard Van Wijk, a Dutch psychologist, wondered how far this influence extended. Did a living thing register information from the entire environment, and not simply between two communicating entities? When a healer sends out healing intention, for instance, how far does his field of influence extend? Would he only affect his target, or would his aim have a shotgun effect, affecting other living organisms around the target?

Van Wijk placed a jar of *Acetabularia acetabulum*, another simple algae, near a healer and his patient, then measured the photon emissions of the algae during healing sessions and periods of rest. After analysing the data, he discovered remarkable alterations in the photon count of the algae. The quality of emissions significantly changed during the healing sessions, as though the algae were being bombarded with light. There also seemed to be changes in the rhythm of the emissions, as though the algae had become attuned to a stronger source of light.

During his initial research, Popp had discovered a strange reaction to light by a living thing. If he shone a bright light on an organism, after a certain delay, the organism would shine more brightly itself with extra photons, as if it were rejecting any excess. Popp called this phenomenon 'delayed luminescence', and assumed it was a corrective device to help the organism maintain its level of light at a delicate equilibrium. In Van Wijk's experiment, the photon emissions of algae showed highly significant shifts from normal, when plotted on a graph. Van Wijk had generated some of the first evidence that healing light may affect anything in its path.[22]

Gary Schwartz's associate Melinda Connor then demonstrated that intention has a direct effect on this light. For her study she clipped leaves from geranium plants, carefully matching them in pairs for size, health, placement on the plant and access to light and close to identical photon emissions. She asked each of 20 master energy healers to send intentions to one of each pair of leaves, first to reduce emissions and then to increase them. In 29 of the 38 sessions designed to decrease emissions, the light was significantly lowered in the treatment leaves, and in 22 of the 38 trials intending to increase the light, the healers caused a significantly greater glow.[23]

Sometimes a physical jolt to the system triggers a shock of realization. For physicist Konstantin Korotkov, his insight resulted from a fall off a roof. It was the winter of 1976, and Korotkov, who was 24 at the time, had been celebrating a birthday with some friends.

Korotkov liked to celebrate outside, whatever the weather. He and his friends had been drinking vodka on the roof. Korotkov was given to expansive gestures, and during a moment of gaiety, threw himself off the roof onto what he thought was a deep bed of snow, which he assumed would cushion his fall. But hidden beneath the snow lay hard stone. Korotkov broke his left leg and landed in the hospital for months.[24]

During his long recovery, Korotkov, a conventional professor of quantum physics at St Petersburg State Technical University in Russia, pondered on a lecture on Kirlian effects and healing that he had attended earlier that year. He had been so intrigued that he wondered if he could improve on what Kirlian claimed to be doing: capturing someone's life energy on film.

Semyon Davidovich Kirlian was an engineer who had discovered in 1939 that photographing living things that had been exposed to a pulsed electromagnetic field would capture what many have termed the human 'aura'. When any conductive object (like living tissue) is placed on a plate made of an insulating material, such as glass, and exposed to high-voltage, high-frequency electricity, a low current results that creates a corona discharge, a halo of coloured light around the object that can be captured on film. Kirlian claimed that the state of the aura reflected the person's state of health; changes in the aura were evidence of disease or mental disturbance.

The Soviet scientific mainstream ignored Kirlian until the 1960s, when the Russian press discovered bioelectrography, as it came to be called, and hailed him as a great inventor. Kirlian photography suddenly became respectable, particularly in space research, and was championed by many Western scientists. Publication of Kirlian's first study in 1964 further attracted the scientific community.[25]

Lying for months in his bed, Korotkov realized that if he was going to discover more about how to capture this mysterious light Kirlian claimed was so vital to health, he was going to have to give up his day job. He knew that the involvement of a well-established quantum physicist such as himself would lend the technique scientific legiti-macy and his technical ability might also help advance the technol-ogy. Perhaps he could even devise a means of depicting the light in real time.

After he got back up on his feet, Korotkov spent months develop-ing a mechanism, which he called the Gas Discharge Visualization (GDV) technique, that made use of state-of-the-art optics, digitized television matrices and a powerful computer. Ordinarily, a living thing will dribble out the faintest pulse of photons, perceptible only to the most sensitive equipment in conditions of utter pitch black. As Korotkov realized, a better way to capture this light was to stir up photons by 'evoking', or stimulating them into an excited state so that they would shine millions of times more intensely than normal.

His equipment blended several techniques: photography, measure-ments of light intensity and computerized pattern recognition. Korotkov's camera would take pictures of the field around each of the 10 fingers, one finger at a time. A computer program would then extrapolate from this a real-time image of the 'biofield' surrounding the organism and deduce from it the state of the organism's health.

Korotkov went on to write five books on the human bioenergy field.[26] In time, he managed to convince the Russian Ministry of Health of the importance of his invention to medical technology, diagnosis and treatment. His equipment was initially employed to predict certain clinical situations, such as the progress of recovery of people after surgery.[27] It soon became widely used in Russia as a diag-nostic tool for many illnesses, including cancer and stress,[28] and was even used to assess athletic potential – to predict the psychophysical reserves in athletes training for the Olympics and the likelihood of victory or exhaustion from overtraining.[29] Eventually, some 3000 doctors, practitioners and researchers worldwide came to use the technology. The National Institutes of Health got interested and funded work on the 'biofield', which employed Korotkov's equipment.[30]

While officially exploring these practical applications, Korotkov privately carried on with his own studies of what had really captured his imagination: the connection between biofields and consciousness.[31] He took GDV readings of healers and a *Qigong* master while they were sending energy, and discovered remarkable changes in their corona discharges. Korotkov then explored the effects of a person's thoughts on the people surrounding him. He asked a number of couples to 'send' a variety of thoughts to their partners, while they were standing within close range. Every strong emotion – whether love, hate or anger – produced an extraordinary effect on the light discharge of the recipient.[32]

Some 40 years after Backster first employed his crude polygraph mechanism to register the effect of thoughts, Korotkov verified those early discoveries with state-of-the-art equipment. He hooked up a potted plant to his GDV machine and asked his researchers to think of different emotions – anger, sadness, joy – and then positive and negative intentions towards the plant. Whenever a participant mentally threatened the plant, its energy field diminished. The opposite occurred if people approached the plant with water or feelings of love.

Largely because he lacked scientific credentials, Backster was never recognized for his contributions. He had stumbled across the first evidence that living things engage in a constant two-way flow of information with their environment, enabling them to register even the nuances of human thought. The more advanced scientific knowledge of physicists Fritz Popp and Konstantin Korotkov was needed to uncover the actual mechanism of that communication. Their research into the nature of quantum light emissions from living organisms suddenly made sense of Backster's findings. If thoughts are another stream of photons, it is perfectly plausible that a plant could pick up the signals and be affected by them.

The work of Backster, Popp and Korotkov suggested something profound about the effect of intention. Every last thought appeared to augment or diminish something else's light.

CHAPTER FOUR

Hearts that Beat as One

NONE OF THE SCIENTISTS INVOLVED IN 'The Love Study' remembered who came up with its name. It might have started as Elisabeth Targ's private joke, for the study involved couples who were installed in two different rooms and separated by a hallway, three doors, eight walls and several inches of stainless steel.[1]

The name was actually meant to be a gracious nod to the study's arcane benefactor, the Institute for Research on Unlimited Love at Case Western Reserve. As it happened, the study became a posthumous valentine to Targ, who was diagnosed with a fatal brain tumour just before the grant money came through. The Love Study would be a fitting tribute to Targ, as the first major scientific demonstration of exactly how intention physically affects its recipient, and the name proved especially apt in describing this process. When you send an intention, every major physiological system in your body is mirrored in the body of the receiver. Intention is the perfect manifestation of love. Two bodies become one.

Targ began her career as a mainstream psychiatrist, but made her name in 1999 with two remarkable studies at California Pacific

Medical Center (CPMC) in San Francisco, which tested the possibility of remote healing with end-stage AIDS patients. Targ spent months designing her trial. She and her partner, psychologist and retired hospital administrator Fred Sicher, sought out a homogeneous group of advanced AIDS patients with the same degree of illness, including the same T-cell counts and number of AIDS-defining illnesses. Because they wished to test the effect of distant healing, and not any particular healing modality, they decided to recruit highly experienced, successful healers from diverse backgrounds who might represent an array of approaches.

Targ and Sicher gathered together an eclectic mix of healers from all across America – from orthodox Christians to Native American shamans – and asked them to send healing thoughts to a group of AIDS patients under strict double-blind conditions. All healing was to be done remotely so that nothing, such as the presence of a healer or healing touch, could confound the results. Targ created a strict double-blind rota: each healer received sealed packets with information about the patients to be healed, including their name, photo and T-cell counts. Every other week, the healers were assigned a new patient and asked to hold an intention for the health and well-being of the patient an hour a day for six days, with alternate weeks off for rest. In this manner, eventually every patient in the healing group would be sent healing by every healer in turn.

At the end of the first study, although 40 per cent of the control population died, all 10 of the patients in the treatment group were not only alive but far healthier in every regard.

Targ and Sicher repeated the study, but this time, doubled the size of their study population and tightened their protocol even further. They also widened their brief of the outcomes they planned to measure. In the second study, those sent healing were again far healthier on every parameter tested: significantly fewer AIDS-defining illnesses, improved T-cell levels, fewer hospitalizations, fewer visits to the doctor, fewer new illnesses, less severity of disease and better psychological well-being. The differences were decisive; for instance, the treatment group had six times fewer AIDS-defining illnesses and four times fewer hospitalizations at the end of the study than the controls.[2]

In Targ's original studies, the healing had been carried out by highly experienced, successful healers who had been chosen because they possessed a special gift. After the studies were completed, Targ grew interested in whether an ordinary individual could be similarly trained to use intention effectively.

For the Love Study, Targ found a sympathetic partner in Marilyn Schlitz, the vice president for research and education at the Institute of Noetic Sciences (IONS). The energetic blonde had a colourful national reputation because of her meticulously designed parapsychology studies and their spectacular results, which attracted the attention of the senior powers in consciousness research as well as the *New York Times*. During a long partnership with psychologist William Braud, Schlitz had conducted rigorous research into what became known in the psychic community as 'DMILS' – direct mental interaction with living systems – the ability of human thought to influence the living world around it.[3] Throughout her career in parapsychology, Schlitz had been fascinated by remote influence; she was one of the first to examine the effect of intention in healing, and went on to assemble a vast database of healing research for IONS.

For the Love Study, Schlitz recruited Dean Radin, her IONS senior researcher and one of America's most renowned parapsychologists. Radin was to design both the study and some of its equipment; with his background in engineering and psychology he would ensure that both the study protocol and its technical detail were pristine. Targ enlisted Jerome Stone, a nurse and practising Buddhist who had worked with her on the AIDS studies, to design the intention programme and train the patients.

In 2002, after Targ died, Schlitz and the others vowed to carry on with the study and recruited Ellen Levine, one of Targ's colleagues from CPMC, to take her place and work with Stone as joint principal investigators.

The Love Study was to follow the basic study design of a perennial favourite among consciousness researchers: the sense of being stared at.[4] In those studies, two people are isolated from each other in separate rooms and a video camera is trained on the receiver, who is also hooked up to skin conductance equipment, not unlike a polygraph machine – the type used in lie detection studies to detect an increase in 'fight-or-flight', unconscious autonomic nervous system activity. At random intervals, the 'sender' is instructed to stare at the subject on the monitor, while the 'receiver' is told to relax and try to think of anything other than the prospect of being stared at. A later comparison analysis determines whether the receiver's autonomic system registered a reaction during those moments he or she was being stared at to determine whether the mere attention of the sender was unconsciously picked up by the most automatic systems of the receiver's body.

Schlitz and Braud's body of evidence on remote staring, conducted over 10 years, showed exactly such an effect. All the studies had been combined into a review that was published in a major psychology journal. The review concluded that the effects had been small but significant.[5]

The Love Study's design was also inspired by the major DMILS studies conducted since 1963, which demonstrated that, under many types of circumstances, the electrical signalling in the brains of people gets synchronized.[6] The frequencies, amplitudes and phases of the brain waves start operating in tandem. Although the studies followed slightly different designs, all of them asked the same question: can the stimulation of one person be felt in the higher central nervous system of another? Or, as Radin liked to think of it, after a sender gets pinched, does the receiver also feel the 'ouch'?[7]

Two people wired up with a variety of physiological monitoring equipment, such as EEG machines, were isolated from each other in different rooms. One would be stimulated with something – a picture, a light or a mild electric shock. The researchers would then examine the two EEGs to determine if the receiver's brain waves mirrored those of the sender when he or she was being stimulated.

The earliest DMILS research had been designed by psychologist and consciousness researcher Charles Tart, who carried out a series of brutal studies to determine whether people could empathetically feel another person's pain. He administered shocks to himself, while a volunteer, isolated in a different room and hooked up to an array of medical gadgetry, was being monitored to see if his sympathetic nervous system somehow picked up Tart's reactions. Whenever Tart jolted himself, the receiver registered an unconscious empathetic response in decreased blood volume and increased heart rate – as though he were also getting the shocks.[8] Another fascinating early study had been carried out with identical twins. As soon as one twin closed his eyes and his brain electrical rhythms slowed to alpha waves, the other twin's brain also slowed, even though his eyes were wide open.[9]

Harald Walach, a German scientist at the University of Freiburg, tried an approach that was guaranteed to magnify the sender's effects, in order to maximize the response in the receiver. The sender was shown an alternating black-and-white checkerboard, called a 'pattern reversal', which is known to trigger predictable, high-amplitude electrical brain waves in viewers. At the same instant, the EEG of the distant, shielded receiver recorded identical brain-wave patterns.[10]

Neurophysiologist Jacobo Grinberg-Zylberbaum, of the National Autonomous University of Mexico in Mexico City, had used this same protocol a decade before Walach but with a different twist: with light flashes rather than patterns as the stimulus. In this study, the particular patterns of firing in the brain of the sender, evoked by the light, turned out to be mirrored in the brain of the receiver, who was sitting in an electrically shielded room 14.5 metres away. Grinberg-Zylberbaum also discovered that an important condition determined success: the synchrony only occurred among pairs of participants who had met and established a connection by spending 20 minutes with each other in meditative silence.[11]

In earlier work, Grinberg-Zylberbaum had discovered that brain-wave synchrony occurred not only between two people, but between both hemispheres of the brains of both participants, with one important distinction: the participant with the most cohesive quantum wave patterns sometimes set the tempo and tended to influence the other. The most ordered brain pattern often prevailed.[12]

In the most recent DMILS study, in 2005, a group of researchers from Bastyr University and the University of Washington gathered 30 couples with strong emotional and psychological connections and also a great deal of experience in meditation. The pairs were split up and placed in rooms 10 metres away from each other, with an EEG amplifier wired up to the occipital (visual) lobe of the brain of each participant. The moment each sender was exposed to a flickering light, he attempted to transmit an image or thought about the light to the partner. Of the 60 receivers tested, 5 of them, or 8 per cent, were shown to have significantly higher brain activation during times their partner 'sent' their visual images.[13]

The Washington researchers then selected five pairs of the participants who had scored a significant result, wired them up to a functional MRI, which measures minuscule changes in the brain during critical functions, and asked them to repeat the experiment. During the times the thought was 'transmitted', the recipients experienced an increase in blood oxygenation in a portion of the visual cortex of the brain. This increase did not occur when the sending partner was not being visually stimulated.[14] The Bastyr researchers replicated their study, this time with volunteers highly experienced in meditation, and got some of the strongest correlations between senders and receivers of all the studies thus far.

The Bastyr study represented a major breakthrough in research on direct mental influence. It demonstrated that the brain-wave response of the sender to the stimulus is mirrored in the receiver, and that the stimulus in the receiver occurs in an identical place in the brain as that of the sender. *The receiver's brain reacts as though he or she is seeing the same image at the same time.*

A final extraordinary study examined the effect of powerful emotional involvement on remote influence. Researchers at the University of Edinburgh studied and compared the EEGs of bonded couples, matched pairs of strangers, and several individuals with no partner but who nevertheless thought they were being paired off and having their brain waves compared. Everyone who had been paired off, whether he knew his partner or not, displayed increased numbers of brain waves in synchrony. The only participants who did not demonstrate this effect were those who had no partner.[15]

Radin carried out a variation of this experiment, attaching pairs who had close bonds – couples, friends, parents and their children. In a significant number of instances, the EEGs of the senders and receivers appeared to synchronize.[16]

In designing the Love Study, Schlitz and Radin also had been influenced by other research showing that, during acts of remote influence, the recipient's EEG waves mirror those of the sender. In a number of studies of healing, the EEG waves of the patient synchronize with those of the healer during moments when healing energy is being 'sent'.[17] Brain mapping during certain types of healing, such as bioenergy, also shows evidence of brain-wave synchrony.[18] In many instances, when one person is sending focused intention to another, their brains appear to become entrained.

Entrainment is a term in physics which means that two oscillating systems fall into synchrony. It was coined in 1665 by the Dutch mathematician Christiaan Huygens, after discovering that two of his clocks with pendulums standing in close approximation to each other had begun to swing in unison. He had been toying with the two pendulums and found that even if he started one pendulum swinging at one end, and the other at the opposite end, eventually the two would swing in unison.

Two waves peaking and troughing at the same time, are considered 'in phase', or operating in synch. Those peaking at opposite times are 'out of phase'. Physicists believe that entrainment results from tiny exchanges of energy between two systems that are out of phase, causing one to slow down and the other to accelerate until the two are in phase. It is also related to resonance, or the ability of any system to absorb more energy than normal at a particular frequency (the number of peaks and troughs in one second). Any vibrating thing, including an electromagnetic wave, has its own preferential frequencies, called 'resonant frequencies', where it finds vibrating the easiest. When it 'listens' or receives a vibration from somewhere else, it tunes out all pretenders and only tunes into its own resonant frequency. It is a bit like a mother instantly recognizing her child from among a mass of school children. Planets have orbital resonances. Our sense of hearing operates through a form of entrainment: different parts of a membrane of the inner ear resonate to different frequencies of sound. Resonance even occurs in the seas, such as in the tidal resonance of the Bay of Fundy in the northeast end of the Gulf of Maine, near Nova Scotia.

Once they march to the same rhythm, things that are entrained send out a stronger signal than they do individually. This most commonly occurs with musical instruments, which sound amplified when all playing in phase. At the Bay of Fundy, the time required for a single wave to travel from the bay's mouth to its opposite end and back is exactly matched by the time of each tide. Each wave is amplified by the rhythm of each tide, resulting in some of the highest tides in the world.

Entrainment also occurs when someone sends a strong intention to cause harm, which became evident in the *tohate* experiments of Mikio Yamamoto of the National Institute of Radiological Sciences in Chiba and the Nippon Medical School in Tokyo. *Tohate* is a kind of mental stand-off between two *Qigong* practitioners, one of whom receives a sensory shock and is eventually made to submit and move back several yards without any physical contact from the other. The central question posed by the technique, in Yamamoto's mind, was whether the effect of *tohate* is psychological or physical: does the opponent move back because of psychological intimidation, or is he knocked over by the *qi* of his opponent?

In the first of Yamamoto's studies, a *Qigong* master was isolated in an electromagnetically shielded room on the fourth floor of a building, while his student was similarly isolated on the first floor. Yamamoto signalled for the master to perform '*qi* emission' over 80 seconds at random intervals. Each time, he tracked their separate movements – the sending of the *qi* and the start of the pupil's recoil. In nearly a third of the 49 such trials – a highly significant result – whenever the master engaged in *tohate* movements, his opponent in the other room was physically knocked back. In a second set of 57 trials, Yamamoto wired both teacher and pupil to EEG machines. Whenever the master emitted *qi*, his pupil showed an increase in the number of alpha brain waves in his right frontal lobe, suggesting that this was where the body initially receives the intention 'message'.

Yamamoto's final set of trials examined the EEG-recorded brain waves of both master and student. Whenever the master performed *tohate*, the beta brain waves of both men demonstrated a greater sense of coherence.[19] In an earlier study carried out by the Tokyo group, the brain waves of the receiver and sender became synchronized within one second during *tohate*.[20]

Besides resonance, the DMILS studies offered evidence of another phenomenon during intention: the receiver anticipated the information by registering the 'ouch' a few moments *before* the pinch occurred in the sender. In 1997, in his former laboratory at the University of Nevada, Radin discovered that humans may receive a physical foreboding of an event. He set up a computer that would randomly select photos designed to calm, to arouse, or to upset a participant. His volunteers were wired to physiological monitors that recorded changes in skin conduction, heart rate and blood pressure, and they sat in front of a computer that would randomly display colour photos of tranquil scenes (landscapes), or scenes designed to shock (autopsies) or to arouse (erotic materials).

Radin discovered that his subjects were registering physiological responses *before* they saw the photo. As if trying to brace themselves, their responses were highest before they saw an image that was erotic or disturbing. This offered the first laboratory proof that our bodies unconsciously anticipate and act out our own future emotional states and that the nervous system does not merely cushion itself against a future blow, but also works out the emotional meaning of it.[21]

Dr Rollin McCraty, executive vice-president and director of research for the Institute of HeartMath, in Boulder Creek, California was fascinated by the idea of shared physical foreboding of an event, but wondered where exactly in the body this intuitive information might first be felt. He used the original design of Radin's study with a computerized system of randomly generated arousing photos, but hooked up his participants to a greater complement of medical equipment.

McCraty discovered that these forebodings of good and bad news were felt in both the heart and brain, whose electromagnetic waves would speed up or slow down just before a disturbing or tranquil picture was shown. Furthermore, all four lobes of the cerebral cortex appeared to take part in this intuitive awareness. Most astonishing of all, the heart appeared to receive this information moments *before* the brain did. This suggested that the body has certain perceptual apparatus that enables it continually to scan and intuit the future, but that the heart may hold the largest antenna. After the heart receives the information, it communicates this information to the brain.

McCraty's study had shown certain fascinating differences between the sexes. Both the heart and brain became entrained with each other earlier and more frequently in women than they did in men. McCraty concluded that this offered scientific evidence of the universal assumption that women are naturally more intuitive than men and more in touch with their heart centre.[22]

McCraty's conclusion – that the heart is the largest 'brain' of the body – has now gained credibility after research findings by Dr John Andrew Armour at the University of Montreal and the Hôpital du Sacré-Coeur in Montreal. Armour discovered neurotransmitters in the heart that signal and influence aspects of higher thought in the brain.[23] McCraty discovered that touch and even mentally focusing on the heart cause brain-wave entrainment between people. When two people touched while focusing loving thoughts on their hearts, the more 'coherent' heart rhythms of the two began to entrain the brain of the other.[24]

Armed with this new evidence about the heart, Dean Radin and Marilyn Schlitz decided to explore whether remote mental influence extended to anywhere else in the body. An obvious place to explore was the gut. People speak about intuition as a 'gut instinct' or 'gut feeling'. Certain researchers have even referred to the gut as a 'second brain'.[25] Radin wondered if a gut instinct was accompanied by an actual physical effect.

Radin and Schlitz gathered 26 student volunteers, paired them, and this time wired them up to an electrogastrogram (EGG), which measures the electrical behaviour of the gut; monitors on the skin usually closely match the frequencies and contractions of the stomach. Although the Freiburg study had shown otherwise, Radin and Schlitz believed that familiarity could only help to magnify the effects of remote influence. In case some sort of physical connection was indeed important, Radin asked all the participants to exchange some meaningful object first.

Radin put one participant from a pair in one room. The other sat in another, darkened room, attached to an electrogastrogram, viewing live video images of the first person. Images periodically flashed on another monitor, accompanied by music designed to arouse particular emotions: positive, negative, angry, calming or just neutral.

The results revealed another example of entrainment – this time in the gut. The EGG readings of the receiver were significantly higher and correlated with those of the sender when the sender experienced strong emotions, positive or negative. Here was yet more evidence that the emotional state of others is registered in the body of the receiver – in this case, deep in the intestines – and that the home of the gut instinct is indeed the gut itself.[26]

This latest evidence was further proof that our emotional responses are constantly being picked up and echoed in those closest to us.[27] In every one of these studies, the bodies of the pairs had become entrained or 'entangled' as Radin called it;[28] the recipients were 'seeing' or feeling what their partners actually saw or felt, in real time.

As this research intimates, intention might be an attunement of energy. The DMILS research established that, under certain conditions, the heart rate, the arousal of the autonomic nervous system, the brain waves and the blood flow to the extremities of different people all become entrained, even when they are situated at a distance. Nevertheless, in most of the DMILS studies, the correlated response resulted from a simple stimulation of the sender, which the recipient unconsciously picked up. Except for one instance, no one attempted to influence another person.

Schlitz and Radin now wanted to find out whether they would achieve similar correlations if the sender were actually sending an intention to heal. For the Love Study, Schlitz and her colleagues decided to recruit ordinary individuals and train them in healing techniques. They wondered whether certain conditions were more favourable than others for achieving entrainment. Many healing studies intimated that motivation, interpersonal connection and a shared belief system were vital to success. Grinberg-Zylberbaum believed that a 'transferred potential', as he termed this form of entrainment, occurred only among those who had undergone some meditative regime and then only after some sort of psychic connection between sender and receiver had been established. Nevertheless, in the Freiberg study, many of the pairs had never met each other and had not had a chance to establish a bond. The German researchers had concluded that 'connectedness' and mental preparation may play a role, but were not crucial. In Schlitz's view, motivation was a key component of success. The more urgent the situation, such as would occur with a partner suffering from cancer, the more motivated his or her partner would be in attempting to get him or her well.

Schlitz and her fellow researchers decided to seek out couples with a wife suffering from breast cancer, and began advertising around the San Francisco Bay Area for volunteers. It soon became apparent that they would have to widen their original brief. The breast-cancer population of the Bay Area, which is higher than average in the USA, has been extremely well studied. From the lacklustre response to their advertising, it appeared that sufferers were unwilling to take part in yet more research. The scientists decided to open the study to any couple if either partner were suffering from cancer of any variety. Eventually 31 couples volunteered, including healthy couples who were to act as controls.

Jerome Stone wrote a training manual for the couples, after analysing a number of healers and distilling their common practices.[29] The first component of his programme involved teaching the sender how to focus and concentrate, as occurs in meditation, to create a high degree of sustained *attention*. The scientific evidence demonstrates that meditation establishes more coherent brain waves; at least 25 studies show that EEG synchronization occurs between the four regions of the brain during meditation.[30] Other studies of meditation have shown that it creates more coherent biophoton emissions[31] and in general aids healing.

Stone also believed that his senders needed to learn how to generate *compassion* or empathy for their partners, with a technique based largely on the Tonglen Buddhist idea of 'giving and receiving'. This practice would train the partner to develop a true understanding of the suffering of another, to take on the suffering without being burdened by it, and to transform it through the process of sending healing. Developing true empathy would also help to dissolve the boundaries and sense of self between the sender and receiver. Positive, loving thoughts also had positive physiological effects. Rollin McCraty's research at HeartMath showed that a steady (or, as they called it, 'coherent') variation in heartbeat was more likely with 'positive' – loving or altruistic – thoughts and that this 'coherence' was quickly picked up by the brain, which soon pulsed in synchrony[32] and evidenced improved cognitive performance.[33]

After Stone instructed the partners in simple techniques of meditation, he also taught them to be compassionate when carrying out intention. The final aspect of Stone's training involved instilling *belief* and confidence in both senders and receivers. Stone had discovered evidence in both the healing and parapsychological literature that belief in the process assists in the success of psychic processes such as ESP, which, like intention, involves 'transferring' information across distance.[34]

Although the training programme was originally intended to run for eight weeks, limited funding meant that Stone had to compress his workshop into a single day, to be followed up with homework and practice.

Radin divided the couples into three groups. The first group (the 'trained group') was to undergo Stone's training, practise compassionate intention daily for three months and then carry out the test. The second group (called the 'wait group') was to carry out the test first and then have the training. The 18 healthy couples comprising the third group (the control group) was to have no training at all, but simply undergo the test.

With all three groups, the member of the couple with the cancer (or one of the designated partners in the control group) was asked to sit in a black reclining chair placed in a one-ton, solid steel, double-walled, electromagnetically shielded enclosure. The tiny Lindgren/ETS chamber was separated from the outside world by two layers of steel and one of solid wood, which blocked out all sound and all electromagnetic energy. Any electrical signals were carried out of the chamber by a fibre-optic cable, to ensure that the room remained, electromagnetically speaking, a solitary confinement.

Each inhabitant was fitted to an array of medical gadgetry to measure brain waves, heartbeat, breathing rate, skin conductance and peripheral blood flow. A video camera stood discretely in the corner.

The room was curtained in earth tones and furnished with soft table lighting and an artificial, floor-to-ceiling weeping fig tree. When the room was occupied, ambient music flooded the space. The furnishings and music, and even a large colour poster of a cascading mountain stream, were all intended to distract from the fact that once the 400-pound steel door with an articulated closing mechanism snapped shut, the inhabitant was essentially trapped inside the warmer equivalent of a meatpacking-plant refrigerator.

Some 20 metres away, the other partner was seated in the dark, attached to the same medical equipment as his or her partner, staring at a small blank TV screen. Bunched towels blocked out the last vestiges of light. Whenever the image of the partner in the refrigerator room abruptly flashed on the television screen, the other member of the couple was to send a compassionate intention to his or her partner for 10 seconds.

Stone, Radin and their colleagues planned to examine two different outcomes: whether the training improved the marriage, and also whether there was any correspondence between the physical sensations of sender and receiver. Although they hoped to examine whether the intentions sent also affected the medical prognosis, limited funding made that aspect of the study impossible.

Stone and Levine were given the task of analysing the social aspects of the study. Initially they discovered that the training made no difference to the quality of the couples' marriages. The finding was not altogether surprising, considering that anyone prepared to be part of a study involving three months of training was already likely to be extremely committed to the partnership. And Schlitz had aimed to recruit motivated partners when she designed the study. A later, more detailed analysis of the figures showed that the intention training and practice had indeed improved the couples' marriages, but Radin concluded that these effects were due to their expectation of improved relations.

Then Radin compiled all the physiological data from the three groups and studied the results between partners and group composite averages. Each physiological response offered fascinating information about the effect of intention on the receiver. For instance, in the case of measurements of blood to the extremities, in every group, the sender's skin conductance increased 2 seconds after seeing the partner's image, and the receiver recorded a similar arousal a half second after the image had flashed. However, unlike the earlier DMILS studies, where the skin-conduction response in the receiver resembled that of a 'startle reflex' and quickly tailed off, in this instance the response persisted 7 seconds after the stimulus. The receiver clearly appeared to be responding to intention – indeed, almost instantaneously. In fact, the receiver's response occurred at least 1 second faster than it would have been possible for the sender to have consciously formulated an intention. Radin was not sure whether this meant that the receiver had had a premonition of the intention. It might simply have reflected the turgid nature of the skin conductance response; the receiver was likely responding in his or her extremities to information sent by the sender's central nervous system, which would have reacted to the initial stimulation of the image on the monitor far more quickly than the electrical impulses sent to his or her fingertips. Nevertheless, in Radin's view, the two skin conductance responses were tracking each other, even if they were slightly out of phase.

A similar situation occurred with the heart rate. The sender's heart rate increased 5 seconds after the stimulus prompt to send the intention – which was consistent with the physical response that occurs in the body during the process of making some sort of mental effort. But an identical increase took place in the receiver, which would not happen ordinarily if he or she were simply resting in a recliner.

Blood flow followed a similar pattern. Whenever we experience something that stimulates us, the vascular network in our extremities constricts slightly, to maximize blood flow to the core of the body. In the Love Study, this phenomenon occurred in the sender, and was soon imitated in the body of the receiver.

As for respiration, on average, whenever the stimulus image appeared, the sender immediately inhaled sharply and blew out the air 15 seconds later. This respiratory response resembles that of someone about to steady himself for the task at hand. In this case, Radin witnessed a different response in the receiver. During the first 5 seconds, the receiver's respiration faltered, almost as though he or she had stopped breathing, and then resumed with a large exhale in the final 5 seconds of the intention. It was as though the receiver had been listening with care, holding her breath and straining to hear something, before sighing with relief as soon as the stimulation had passed.

But it was the brain-wave results that proved to be the most interesting. Whenever the receiver's image flashed on the screen, the senders recorded a little upturn in brain waves, like a 'flinch response', and then a huge spike for about a third of a second before they dropped sharply and took about one second to come back to baseline. In the sender, this tiny initial upturn represents something called a P300 wave – a well-established phenomenon that records the time that the brain takes to process the switching on of a light. The drop represents the time it takes for internal attention to modulate the stimulus into a response.

In this instance, the receivers had no P300 wave, but their brain waves nevertheless mimicked the virtually vertical plunge of the brain wave that shortly followed in the sender, even though, unlike the sender, the receiver had had no stimulus. The brain of the receiver was reacting just as it does when asleep and dreaming. The receivers had registered an emotional reaction, even though there was no tangible stimulus.

Radin's results were all the more remarkable because the receivers had not been told how long the stimulus period would be, and neither senders nor receivers knew in advance how long the sender would have to wait before the partner's image flashed on screen. A computer program randomly selected the time frame, which ranged from 5 to 40 seconds. This meant that any expectation on the part of either member of the couples could not explain the results.

Radin then compared the responses of the groups. All three groups had shown an effect. In every instance, each physiological response of the receivers had tracked those of the senders. However, the most prolonged pattern occurred among the cancer patients whose partners had been trained in compassionate intention. The receivers in the training group not only responded to the stimulus, but also kept responding over 8 of the 10 seconds of the intention. In quantum terms, the couples had become as one.[35]

The Love Study indicates a number of profound suggestions about the nature of intention. Sending a directed thought seems to generate a palpable energy; whenever one of Radin's senders sent a healing intention, many subtle aspects of the receiver's body became activated, as though he had received a minuscule electric shock. It seemed to be a kind of activating awareness, as though his body had *felt or heard* the healing signal.

There had even been an element of anticipation in the receiver; some of the physiological reactions recorded suggested that the receiver had felt the partner's healing intention *before* he had even sent it.

People appear to receive healing deep in their bodies by being retuned to the more coherent energy of the healer's intention. During healing, it could be that the 'orderly' energy of the well person entrains and 're-orders' the sick.

In order to have the most powerful effect, a healer or sender needs to become 'ordered' on some subatomic level, mentally and emotionally. The Love Study demonstrates that certain conditions and mental states make our intention especially powerful and ourselves more ordered, and that these states can be achieved with training. The success of the basic training programme that Schlitz, Radin and Stone assembled suggests that attention, belief, motivation and compassion are important for intention to work, but there are probably other conditions that intensify its effects.

I needed, for instance, to find out how we can loosen our psychological boundaries. It was becoming clear to me: when we send intention, in a manner of speaking, we have to 'become' the other.

PART TWO

Powering Up

For every atom belonging to me
as good belongs to you.
'Song of Myself', Walt Whitman

CHAPTER FIVE

Entering Hyperspace

IN A DRAUGHTY MONASTERY high in the Himalayas in northern India, during the winter of 1985, a group of Tibetan Buddhist monks were seated quietly, deep in meditation. Although scantily clad, they appeared oblivious to the chilly indoor air temperature, which approached freezing. A fellow monk passed between them, draping each, in turn, with sheets drenched with cold water. Such extreme conditions would ordinarily shock the body and send the core temperature plummeting. If body temperature falls by only 7°C, within minutes a person will lose consciousness and all vital signs.

Instead of shivering, the monks began to sweat. Steam rose from the wet sheets; within an hour, they were thoroughly dry. The attendant replaced the dry sheets with new ones, also drenched in ice-cold water. By this time, the monks' bodies had become the equivalent of a furnace. Those sheets were efficiently dried, as was a third batch.

A team of scientists led by Herbert Benson, a cardiologist at Harvard Medical School, stood nearby, examining an array of medical equipment to which they had attached the monks for any clues as to what particular physiological mechanism might have enabled the body to generate this extraordinary level of heat. For a number of years, Benson had explored the effects of meditation on the brain and the rest of the body. He'd embarked on an ambitious research programme, studying Buddhists in various remote outposts around the world who had spent many years in disciplined practice. During one trip to the Himalayas, he also videotaped monks, dressed only in light shawls, as they spent a freezing February night outdoors on a mountain ledge 4600 metres above sea level. Benson's film showed that they had slept soundly through the night, without clothing or shelter.

In his travels, Benson had witnessed many extraordinary feats of intention – mastery over temperature or metabolic rate that could even produce a state resembling hibernation. The monks monitored by Benson's team had raised the temperature of their extremities by up to 9.4°C and lowered their metabolism by more than 60 per cent.[1] Benson realized that this represented the largest variation in resting metabolism ever reported. During sleep, by contrast, metabolism only drops by 10 to 15 per cent; even experienced meditators can only decrease it by 17 per cent, at best. But that day in the Himalayas, he had observed the impossible in terms of mental influence. The monks had used their bodies to boil freezing water simply through the power of their thoughts.[2]

Benson's enduring enthusiasm for meditation ignited interest at major academic institutions across America. By the end of the twentieth century, monks had become the favourite guinea pigs of the neuroscience laboratory. Scientists from Princeton, Harvard, the University of Wisconsin and the University of California–Davis followed Benson's lead by wiring up monks to state-of-the-art monitoring equipment and studying the effects of intensive, advanced meditation. Entire conferences were held on meditation and the brain.[3]

It was not the practice itself that fascinated these scientists, but its effect on the human body, particularly the brain, and the possibilities this suggested. By studying the biological effects in such detail, scientists hoped to understand the neurological processes that occur during feats of highly directed thought, as the monks had displayed in the Himalayas.

Monks also offered scientists an opportunity to study whether years of focused attention stretch the brain beyond its usual limits. Did the brain of a monk become the equivalent of an Olympic athlete's body – more highly developed and ultimately transformed after gruelling discipline and practice? Do training and experience change the physiology of the brain over time? Would practice enable you to become a bigger and better transmitter of intention? The answers would in turn address a long-standing debate in neuroscience: is neural structure basically hard-wired from youth or plastic – changeable – depending on the nature of a person's thoughts through life?

For me, the most intriguing question about this research on focused attention was the means by which a Buddhist monk could turn himself into a human boiler, and how these means compared with techniques and practices of other ancient traditions. Like Benson, I was intrigued by 'masters' of intention: practitioners of ancient disciplines – Buddhism, *Qigong*, shamanism, traditional native healing – who had been trained to perform extraordinary acts through their thoughts. I wanted to work out the common denominators they shared. Do the steps taken by a *Qigong* master to send *Qi* resemble those of a Buddhist monk during meditation? Which mental disciplines ensure that a healer will enter a state enabling him to repair another person's body? Are 'masters' of intention graced with special neurological gifts that enable them to use their minds more powerfully than the rest of us, or did they acquire a skill that ordinary people could learn as well? And, perhaps most important, what did the neurological study of monks tell me about the effect of focused intention on the brain? Would practice enable you to become a bigger and better transmitter of intention?

I began studying scientific research about healing methods from a variety of traditions and then conducted my own questionnaire and interviews with healers and 'master' intenders of all persuasions.[4] I was aided in my research by the work of psychologist Stanley Krippner and his student Allan Cooperstein at Saybrook Graduate School. A clinical and forensic psychologist, Cooperstein had conducted a thorough study of the various techniques used by distant healers for his doctoral thesis, including an analysis of scholarly books on healing and exhaustive written and verbal interviews with well-known practitioners who had scientific evidence of success in healing.[5]

In every instance, I discovered, the most important first step involved achieving a state of concentrated focus, or *peak attention*.

According to Krippner, an expert on shamanic and other native traditions, virtually all native cultures carry out remote healing during an altered state of consciousness and achieve a state of concentrated focus through a variety of means.[6] Although the use of hallucinogenic drugs such as ayahuasca is common, many cultures use a strong repetitive rhythm or beat to create that state; the Native American Ojibway *wanbeno*, for instance, use drumming, rattling, chanting, naked dancing and handling of live coals.[7] Drumming is particularly effective in producing a highly concentrated focus; a number of studies have shown that listening to the beat of a drum causes the brain to slow down into a trancelike state.[8] As Native Americans discovered, even intense heat, as in a sweat lodge, can transport individuals to an altered state.

In my own study of intention 'masters', I spoke with Bruce Frantzis, arguably the greatest *Qigong* master in the West. A martial arts champion, with black belts in five Japanese martial arts, he also learned healing *Qigong* through years of study with Chinese masters. Frantzis's powers of intention were legendary; he had been videoed sending people flying across the room simply by directing *Qi*. In his fighting days, he had put several people into wheelchairs. Now, knowing its extraordinary power, he reserved *Qi* for healing. During my own meeting with him, Frantzis gave a short demonstration of the power of directed *Qi*. After a moment of intense concentration, the plates of his skull began to undulate over the top of his head like a rolling surf.[9]

Frantzis taught his students how to develop peak attention gradually, through intense concentration on their breathing. Although they began with very short bursts of 'longevity' breathing, they would work on extending these periods until eventually they could hold this focus continuously. They would also be taught methods of becoming acutely aware of all physical sensation.[10]

The healers I interviewed entered this focused state through a variety of means: meditation; prayer; intense attention on the person to be healed; symbolic or mythic ideas; strong mental images of a situation producing the desired change; verbal affirmations; mental imagery; even internal autosuggestions as a warm-up exercise. One healer established focused attention by saturating his awareness with the goal that he was trying to achieve.

Dr Janet Piedilato, a shamanic healer, will often 'gently hum or chant' or use a 'rattle or other instrument'. Dr Constance Johnson, a Reiki practitioner, can return to an altered state at will. Others need to work hard to achieve this transformation: The Reverend Francis Geddes, a spiritual healer, will meditate on a small object like a pebble, leaf, or twig in a 'very concentrated manner for ten minutes'. Still others use the patient as the object of meditation. As Dr Judith Swack, a mind–body healer who has developed her own holistic psychotherapy system, says: 'I look directly at the client and focus all of my senses forward toward the client and enter a receptive state where I pay internal attention to any subtle information and impression coming in like a kind of radar.' Many other healers likewise enter an altered state, simply by 'listening to the patient' – 'audibly or otherwise'. 'Just thinking of the need to help someone,' wrote Dr Piedilato, 'slows the blood in my veins.'

Initially, many healers experience a heightening of their cognitive processes, but most soon reach a point when inner chatter ceases, and they experience a falling away of all sensation but pure image. The focusing seems to dissolve their own boundaries. They suddenly become aware of the inner workings of the patient's body and ultimately have a sense of being engulfed by the healee.

I was especially interested in the effect of this intense concentration on the activity of the brain. Does the brain slow down or speed up? The received wisdom is that during meditation the brain slows down. The bulk of the research examining the electrical activity of the brain during meditation indicates that meditation leads to a predominance of either alpha rhythms (slow, high-amplitude brain waves with frequencies of 8–13 hertz, or cycles per second), which also occurs during light dreaming, or even the slower theta waves (4–7 hertz), which typify the state of consciousness during deep sleep.[11] During ordinary waking consciousness, the brain operates much faster, using beta waves (around 13–40 hertz). For decades, the prevailing view has been that the optimum state for manifesting intention is an 'alpha' state.

Richard Davidson, a neuroscientist and psychologist at the University of Wisconsin's Laboratory for Affective Neuroscience, recently put this view to the test. Davidson was an expert in 'affective processing' – the place where the brain processes emotion and the resulting communication between the brain and body. His work had come to the attention of the Dalai Lama, who invited him to visit Dharamsala, India, in 1992; a science buff, his Holiness wished to understand more about the biological effects of intensive meditation. Afterwards, eight of the Dalai Lama's most seasoned practitioners of Nyingmapa and Kagyupa meditation were flown to Davidson's lab in Wisconsin. There, Davidson attached 256 EEG sensors to each monk's scalp in order to record electrical activity from a large number of different areas in the brain. The monks were then asked to carry out compassionate meditation. As with Jerome Stone's intention regime, the meditation entailed focusing on an utter readiness to help others and a desire for all living things to be free of suffering. For the control group, Davidson enlisted a group of undergraduates who had never practised meditation and arranged for them to undergo a week's training, then attached them to the same number of EEG sensors to monitor their brains during meditation.

After 15 seconds, according to the EEG readings, the monks' brains did not slow down; they began speeding up. In fact, they were activated on a scale neither Davidson nor any other scientist had ever seen. The monitors showed sustained bursts of high gamma-band activity – rapid cycles of 25–70 hertz. The monks had rapidly shifted from a high concentration of beta waves to a preponderance of alpha, back up to beta and finally up to gamma. Gamma band, the highest rate of brain-wave frequencies, is employed by the brain when it is working its hardest: at a state of rapt attention, when sifting through working memory, during deep levels of learning, in the midst of great flashes of insight.

As Davidson discovered, when the brain operates at these extremely fast frequencies, the phases of brain waves (their times of peaking and troughing) all over the brain begin to operate in synchrony. This type of synchronization is considered crucial for achieving heightened awareness.[12] The gamma state is even believed to cause changes in the brain's synapses – the junctions over which electrical impulses leap to send a message to a neuron, muscle or gland.[13]

That the monks could achieve this state so rapidly suggested that their neural processing had been permanently altered by years of intensive meditation. Although the monks were middle-aged, their brain waves were far more coherent and organized than those of the robust young controls. Even during their resting state, the Buddhists showed evidence of a high ratio of gamma-band activity, compared with that of the neophyte meditators.

Davidson's study bolstered other pieces of preliminary research suggesting that certain advanced and highly focused forms of meditation produce a brain operating at peak intensity.[14] Studies of yogis have shown that, during deep meditation, their brains produce bursts of high-frequency beta or gamma waves, which often are associated with moments of ecstasy or intense concentration.[15] Those who can withdraw from external stimuli and completely focus their attention inward appear more likely to reach gamma-wave hyper-space. During peak attention of this nature, the heart rate also accelerates.[16] Similar types of effects have been recorded during prayer. A study monitoring the brain waves of six Protestants during prayer found an increase in brain-wave speed during moments of the most intense concentration.[17]

Different forms of meditation may produce strikingly different brain waves. For instance, yogis strive for *anuraga*, or a sense of constant fresh perception; Zen Buddhists aim to eliminate their response to the outer world. Studies comparing the two find that the former produces heightened perceptual awareness – magnified outer focus – while the latter produces heightened inner absorption – magnified inner awareness.[18] Most research on meditation has concerned the type that focuses on one particular stimulus, such as the breath or a sound, like a mantra. In Davidson's study, the monks concentrated on having a sense of compassion for all living things. It may be that compassionate intention – and other similar, 'expansive' concepts – produces thoughts that send the brain soaring into a supercharged state of heightened perception.

When Davidson and his colleague Antoine Lutz wrote up their study, they realized that they were reporting the highest measures of gamma activity ever recorded among people who were not insane.[19] In their results they noticed an association between level of experience and ability to sustain this extraordinarily high brain activity; those monks who had been performing meditation the longest recorded the highest levels of gamma activity. The heightened state also produced permanent emotional improvement, by activating the left anterior portion of the brain – the portion most associated with joy. The monks had conditioned their brains to tune into happiness most of the time.

In later research, Davidson demonstrated that meditation alters brain-wave patterns, even among new practitioners. Neophytes who had practised mindfulness meditation for only eight weeks showed increased activation of the 'happy-thoughts' part of the brain and enhanced immune function.[20]

In the past, neuroscientists imagined the brain as something akin to a complex computer, which got fully constructed in adolescence. Davidson's results supported more recent evidence that the 'hard-wired' brain theory was outdated. The brain appeared to revise itself throughout life, depending on the nature of its thoughts. Certain sustained thoughts produced measurable physical differences and changed its structure. Form followed function; consciousness helped to form the brain.

Besides speeding up, brain waves also synchronize during meditation and healing. In fieldwork with indigenous and spiritual healers in five continents, Krippner suspected that, prior to healing, the healers all underwent brain 'discharge patterns' that produce a coherence and synchronization of the two hemispheres of the brain, and integrate the limbic (the lower emotional centre) with the cortical systems (the seat of higher reasoning).[21] At least 25 studies of meditation have shown that, during meditation, EEG activity between the four regions of the brain synchronizes.[22] Meditation makes the brain permanently more coherent – as might prayer. A study at the University of Pavia in Italy and the John Radcliffe Hospital in Oxford showed that saying the rosary had the same effect on the body as reciting a mantra. Both were able to create a 'striking, powerful, and synchronous increase' in cardiovascular rhythms when recited six times a minute.[23]

Another important effect of concentrated focus is the integration of both left and right hemispheres. Until recently, scientists believed that the two sides of the brain work more or less independently. The left side was depicted as the 'accountant', responsible for logical, analytical, linear thinking, and speech, and the right side, as the 'artist', providing spatial orientation, musical and artistic ability, and intuition. But Peter Fenwick, consultant neuropsychiatrist at the Radcliffe Infirmary in Oxford, St Thomas' Hospital, Bethlehem Hospital and the Institute of Psychiatry at the Maudsley Hospital, gathered evidence to show that speech and many other functions are produced in both sides of the brain and that the brain works best when it can operate as a totality. During meditation, both sides communicate in a particularly harmonious manner.[24]

Concentrated attention appears to enlarge certain mechanisms of perception, while tuning out 'noise'. Daniel Goleman, author of *Emotional Intelligence*,[25] carried out research showing that the cortices of meditators 'speed up', but get cut off from the limbic emotional centre. With practice, he concluded, anyone can carry out this 'switching-off' process, enabling the single mode of the brain to experience heightened perception without an overlay of emotion or meaning.[26] During this process, all of the power of the brain is free to focus on a single thought: an awareness of what is happening at the present moment.

Meditation also appears to permanently enhance the brain's reception. In several studies, meditators have been exposed to repetitive stimuli like light flashes or clicks. Ordinarily, a person will get used to the clicks, and the brain, in a sense, will switch off and stop reacting. But the brains of the meditators continued to react to the stimuli – an indication of heightened perception of every moment.[27]

In one study, practitioners of mindfulness meditation – the practice of bringing heightened, non-judgemental awareness of the senses' perceptions to the present moment – were tested for visual sensitivity before and immediately after a three-month retreat, during which time they had practised mindfulness meditation for 16 hours a day. The staff members who did not practise the meditation acted as a control group. The researchers were testing whether the participants could detect the duration of simple light flashes and the correct interval between successive ones. To those without mental training in focusing, these flashes would appear as one unbroken light. After the retreat, the practitioners were able to detect the single-light flashes and to differentiate between successive flashes. Mindfulness meditation enables its practitioners to become aware of unconscious processes and to remain exquisitely sensitive to external stimuli.[28] As these studies indicate, certain types of concentrated focus, like meditation, enlarge the mechanism by which we receive information and clarify the reception. We turn into a larger, more sensitive radio.

In 2000, Sara Lazar, a neuroscientist at Massachusetts General Hospital and an expert in functional magnetic resonance imaging (fMRI), confirmed that this process produces actual physical changes. Conventional MRI employs radio-frequency waves and a powerful magnetic field to view the soft tissues of the body, including the brain. 'Functional' magnetic resonance imaging, on the other hand, measures the minuscule changes in the brain during critical functions. It confirms where and when stimuli and language are being processed by measuring the increase in blood flow in the fine network of arteries and veins of the brain when certain neural networks are engaged. For scientists like Lazar, the fMRI is the closest science can get to observing a brain at work in real time.

Herbert Benson had enlisted Lazar to map the brain regions that are active during simple forms of meditation. Rather than scrutinizing more monks or other meditation 'athletes' who had devoted themselves to the contemplative life, Lazar preferred to study the effect of meditation on the millions of ordinary Americans who performed meditation for just 20–60 minutes a day. She and Benson recruited five volunteers, who had practised Kundalini meditation for at least four years. This kind of meditation employs two different sounds to focus and still the mind while observing inhalation and exhalation of the breath. Lazar asked volunteers to alternate between intervals of meditation and control states, during which they silently ticked off a mental list of animals. Throughout the experiment, Lazar also monitored the biological activity of her subjects – heart rate, breathing, oxygen saturation levels, levels of exhaled CO_2, and EEG levels.

Lazar discovered that, during meditation, the volunteers had a significant increase of signalling in the neural structures of the brain involved in attention: the frontal and parietal cortex, or the 'new' part of the brain where higher cognition takes place, and the amygdala and hypothalamus, portions of the 'old' brain that govern arousal and autonomic control.

This finding was another contradiction of the received wisdom that meditation is always a state of quiescence. Her results offered yet more evidence that, during certain types of meditation, the brain is engaged in a state of rapt attention.

Lazar also discovered that the signalling in certain areas of the brain and the neural activity during meditation evolved over time and increased with meditative experience. Her subjects themselves had the impression that their states of mind continued to change during each individual meditation and as they grew more experienced.[29]

These results suggested to Lazar that highly concentrated focus over time might enlarge certain parts of the brain. To test this, she gathered 20 long-term practitioners of Buddhist mindfulness meditation (five of whom were meditation teachers) with an average of nine years of meditation experience. Fifteen non-meditators acted as controls. Participants meditated in turn inside an ordinary MRI scanner while Lazar took detailed images of their neural structures.

Lazar discovered that those portions of the brain associated with attention, awareness of sensation, sensory stimuli and sensory processing were thicker in the meditators than in the controls. The effects of meditation definitely were 'dose-dependent': increases in cortical thickness were proportional to the overall amount of time the participant had spent meditating.

Lazar's research offered some of the first evidence that meditation causes permanent alterations in brain structure. Up until the time of her experiment, this type of increase in cortical volume had only been linked to certain repetitive mechanical practices requiring a high degree of attention, such as playing an instrument or juggling. Here was some of the first evidence that thinking certain thoughts exercises the 'attention' portion of the brain and makes it grow larger. Indeed, the cortical thickness of these regions was even more pronounced in the older participants. Ordinarily, cortical thickness deteriorates as a result of ageing. Regular meditation appears to reduce or reverse the process.

Besides increasing cognitive processing, meditation also appears to integrate emotional and cognitive processes. In the fMRI study, Lazar found evidence of activation of the limbic brain – the primitive, so-called 'instinctive' part of the brain involved with primitive emotion. Meditation appears to affect not only the brain's reasonable, analytical 'upstairs' but also the unconscious and intuitive 'downstairs'. She had discovered greater activation in the part of the brain responsible for what is usually called 'the gut instinct'. Here was physical evidence that meditation not only increases our ability to receive intuitive information, but also our conscious awareness of it. Davidson had shown increases in the 'approach' portion of the brain – the part that wants to help – in his monks, who were attempting to help humanity by meditating on compassion. They had increased the 'can I help you' portion of their brains. Lazar's meditators, however, were working on mindfulness, a state of peak attention, and that part of the brain responsible for attention had grown larger. The brain's powers of observation had increased, allowing in more information, even the kind that is received intuitively.

Some people are born with a larger-than-normal antenna and better reception than usual. This appears to be the case with the psychic

Ingo Swann. Swann's psychic gifts extended to remote viewing, the ability to perceive objects or events beyond normal human vision. He had helped to develop a remote viewing programme used by the American government and was widely regarded as one of the best remote viewers in the world. Swann once had allowed the peculiar workings of his brain to be monitored and analysed by Michael Persinger, professor of psychology at Laurentian University in Canada. Wired to an EEG machine, Swann was asked to use his skills to identify items in a distant room. At the very moment that he was able to 'see' the items remotely, his brain showed bursts of fast activity in the high beta and gamma range, similar to that of Benson's Tibetan monks. Those bursts of activity occurred primarily over the right occipital region, the portion of the brain relating to sight. According to the results of brain-wave monitoring, Swann had entered a super-conscious state, enabling him to receive information impossible to access during normal waking consciousness.

When examined by MRI, Swann also showed that he had an unusually large parieto-occipital right-hemisphere lobe, the portion of the brain involved with sensory and visual input. Persinger had found a similar neural aberration in another gifted psychic called Sean Harribance.[30] When monitored with EEG and single-photon emission computerized tomography (SPECT) equipment during his psychic activities, Harribance evidenced an increase in firing of the right parietal lobe. Both he and Swann had been graced with a greater capacity than normal to 'see' beyond the limits of time, distance, and the five major senses.

Science has demonstrated that by thinking certain thoughts it is possible for us to alter and enlarge portions of our brains to become a larger, more powerful *receiver*. But is it also possible to develop a larger *transmitter*? To discover some of the qualities that enhance transmission, I would have to study 'masters' of intention who were

particularly gifted at transmitting. The best place to look seemed to
be among talented healers.

Cancer specialist and psychologist Lawrence LeShan, who has
studied how gifted healers work, discovered that they share two
important practices, besides entering an altered state of consciousness: they visualize themselves as uniting with the person to be
healed and imagine themselves and that person as being united with
what they often describe as the absolute.[31]

Cooperstein's healers had also described turning off the ego and
eliminating their sense of self and separateness. They had the sense
of assuming the body and vantage point of the person to be healed.
One healer actually felt his body changing, with shifts of patterns and
distributions of energy. Although the healers did not take on the
disease or pain, they sensed it once they had visualized themselves as
being at one with the person being healed. At this point of union, the
healers' perception markedly altered and their motor skills diminished. They were suffused by an expanded sense of pure present, and
grew unaware of the passage of time. They lost awareness of the
boundaries of their own bodies, and even experienced an altered
sense of bodily image. They felt taller, lighter – almost as though
they were out of their physical being – engulfed by a sense of unconditional love. They began to observe themselves, according to one
healer, only as 'a kind of a core that remains':

> I'm aware of the process *just being beyond me* … My intent is obviously
> with the person – *my conscious control is completely side-stepped*, like I'm
> standing, watching. Then something else takes over … I don't think that
> I ever lose complete awareness that I'm sitting there.[32]

Other healers experienced a more profound loss of identity; to
carry out their work, they had to be at one with the person they
were healing: to *become* that person, complete with his or her physical and emotional history. Their own personal identity and

memory receded and they entered into some space of joint consciousness, where an impersonal self carried out the actual healing. Some of the healers took on a mystical identification with guardian spirits or guides, and the spiritual alter ego took over.

In Krippner's experience, certain personalities are more susceptible to merging identities than others: those who, according to a psychological test, possess 'thin boundaries'. According to the Hartmann Boundary Questionnaire test, a test developed by Tufts University psychiatrist Ernest Hartmann to test a person's psychological armament, people with thick boundaries are well organized, dependable, defensive and, as Hartmann himself liked to put it, 'well armored', with a sturdy sense of self that remains locked around them like a chain-link fence. People with 'thin' boundaries tend to be open, unguarded and undefended.[33] Sensitive, vulnerable and creative, they tend to get involved quickly in relationships, experience altered states, and easily flit between fantasy and reality. Sometimes, they are not sure which state they are in.[34] They do not repress uncomfortable thoughts or separate feelings from thoughts. They tend to be more comfortable than thick-boundaried people with the use of intention to control or change things around them. In a study by Marilyn Schlitz of musicians and artists, for instance, creative individuals with thin boundaries also scored best in remote influence.[35]

Krippner demonstrated the relationship between thin boundaries and intention with students at Ramtha's School of Enlightenment in Yelm, Washington. Many of the techniques taught at the school – for example, focusing on a desired outcome and excluding all external stimuli, blindfolding students and having them find their way around a labyrinth – were designed to help students release their usual boundaries. The school encouraged students to engage in imaginative fantasy, claiming that it opened untapped areas in the brain.[36] Krippner and several colleagues performed psychological tests on six of the long-time students who claimed to have developed keen skills in manifesting intention.

Ian Wickramasekera, a psychologist who participated in some of the Yelm research, had developed a battery of psychological tests based on his High-Risk Model of Threat Perception.[37] Wickramasekera claimed the tests identify people most likely to have psychic experiences or to be susceptible to hypnosis. Although the test was originally developed to pinpoint people at high risk of psychological problems during times of major life changes, Krippner believed Wickramasekera's model could also be used to evaluate mediums and healers. Krippner and his associates found they could readily use the test to identify people whose inflexible sense of reality blocked them from perceiving or acknowledging intuitive information. Wickramasekera's model predicted that individuals would best perform healing if they were able to block the sense of a threat when they let go of their separatist notions of self.

According to their scores, the Ramtha students had extraordinarily thin boundaries. Hartmann's own mean score, derived from tests on 866 individuals, was 273. The Ramtha students scored 343. The only other groups Hartmann had identified with boundaries this thin were music students and people suffering from frequent nightmares. The Ramtha students also showed a high degree of what psychologists call a type of 'dissociation' (the ability to undergo profound disruptions in their attention) and a high degree of absorption (a tendency to lose themselves in ongoing activity such as hypnosis and a readiness to accept other aspects of reality).[38]

In my own examination of healers, I had come across two types. Some regarded themselves as the water (the source of healing); others saw themselves as the hose (the channel for healing energy to travel through). The first group believed the power resulted from their own gift. By far the largest group, however, comprised the channellers – those who acted as vehicles for a greater force beyond themselves.

Elisabeth Targ's AIDS project had recruited 40 healers of every persuasion.[39] Approximately 15 per cent were traditional Christian religious healers, who used the rosary or prayer. Others were members of non-traditional healing schools, such as the Barbara Brennan School of Healing Light, or those taught by Joyce Goodrich or Lawrence LeShan. Some worked on modifying complex energy fields through changing colours or vibrations or the patient's energy field. More than half the healers concentrated on healing a patient's chakras, or energy centres of the body; others worked with tones, reattuning their patients with audible vibrations. A *Qigong* master from China sent harmonizing *Qi* to the patients. One man working in the Native American tradition went into a trance during a traditional drumming and chanting pipe ceremony on the deserted ridges of Chaco Canyon, New Mexico, and claimed to have contacted spirits on behalf of the patients. Much of the imagery the healers used to describe what they did was framed in terms of relaxing, releasing or allowing spirit, light or love in. For some healers, the spirit was Jesus; for others, Starwoman, a healing Native American grandmother image.

Targ had interviewed the healers about their work, and I spoke with her before she died about the common threads she had discovered among their diverse approaches.[40] She found that a quality of loving compassion or kindness was essential in sending out a positive intention to heal. But no matter what their approach, most of them agreed on a single point: *the need to get out of the way.* They surrendered to a healing force. They had framed their intention essentially as a request – *please may this person be healed* – and then stepped back. When Targ examined those patients whose illness had most improved, and analysed which healers they had been exposed to, those healers who were the most successful were the 'channellers' – the ones who had moved aside to allow the greater force in. None of the healers who had been successful believed he possessed the power himself.[41]

Psychiatrist Daniel Benor, who has accumulated and catalogued virtually every study of healing in four volumes[42] as well as on a website,[43] has examined the statements and writings of the most famous healers describing how they work. One of the most remarkable and best-studied healers, Harry Edwards, wrote that a healer worked by handing over his will and his request for healing to a greater power:

> This change may be described (inadequately) as the healer feeling a sense or condition enshrouding him, as if a blind had been drawn over his normal alert mind. In its place he experiences the presence of a new personality – one with an entirely new character – which imbues him with a super-feeling of confidence and power.
>
> ...
>
> [While engaged in his healing] the healer may be only dimly aware of normal movement, speech, etc., taking place around him. If a question is addressed to him about the patients' condition, he will find himself able to respond with extraordinary ease and without mental effort – in other words, the more knowledgeable personality of the Guide provides the answer. Thus does the healer 'tune-in' – it is the subjection of his physical sense to the spirit part of himself, the latter becoming for the time being the superior self under the control of the director.[44]

To Edwards, the most important act was *moving aside*, shedding the personal ego, *making a conscious attempt to get out of the way*.

Cooperstein's healers described their experience as a sense of total surrender to a higher being or even to the process. All believed that they were a part of a larger whole. To gain access to the cosmic, non-local entity of true consciousness, they had to set aside the limiting boundaries of the self and personal identity, and merge with the higher entity. With this change of consciousness and expanded awareness, the healers felt they got onto an open line to this larger information field, which offered them flashes of information, symbols and images. Words would appear, seemingly from nowhere, giving them a diagnosis. Something beyond their conscious thought would carry out the healing for them.

Although the lead-up to healing was accomplished through consciously directed thought, the actual healing often was not. In giving a 2-minute treatment, for instance, they might have a minute and a half of rational thought and then 'a five-second thing that would be an irrational thing, a space that may be the apex, the key to the whole experience'.[45] The most important aspect of the healers' process was undoubtedly their surrender – their willingness to give up their sense of cognitive control of the process and allow themselves to become pure energy.

But was this capacity to *move aside* important in all types of intention? I found an interesting answer in a study of people with brain damage. Investigators at the Behavioural Neurology Program and Rotman Research Institute at the University of Toronto attempted to replicate the work of the Princeton PEAR lab using random-event generators, but with one important twist: they had enlisted several patients with frontal-lobe damage. The patients who had suffered right frontal-lobe damage, which probably affected their ability to focus and maintain attention, had no effect on the machines. The only person to have a greater than normal effect was a volunteer with a damaged left frontal lobe but whose right frontal lobe was intact. The investigators speculated that the volunteer's particular handicap could have given him a reduced sense of self, but a normal state of attention. Achieving a state of a reduced self-awareness – a difficult state for an ordinary person to achieve – might allow for greater effects of intention on the machines.[46]

Krippner suspects that during some altered states of consciousness, the body naturally 'switches off' certain neural connections, including an area near the back of the brain that constantly calculates a person's spatial orientation, the sense of where one's body ends and the external world begins. During a transpersonal or transcendent experience, when this region becomes inactive, the boundary in the relationship between the self and the other blurs; you no longer know where you end and someone else begins.

Eugene d'Aquili, of the University of Pennsylvania, and Andrew Newberg, a medical doctor at the university hospital's nuclear medicine programme, demonstrated this in a study of Tibetan monks. Moments of meditative experience showed up as more activity in the brain's frontal lobes with less activity in the parietal lobes.[47] Meditation and other altered states can also affect the temporal lobes, which house the amygdala, a cluster of cells responsible for the sense of 'I' and our emotional response to the world: whether we like or dislike what we perceive. Stimulation of the temporal lobes or disorder in them may create familiarity or strangeness – common features of a transcendent experience. Intense focus with intention on some other being appears to 'switch off' the amygdala and so remove the neural sense of self.

Davidson, Krippner and Lazar demonstrated that we can remodel particular portions of our own brains, depending on our different types of focus and indeed different thoughts. It became clear to me that the intense focus of certain types of meditation can be a portal to hyperspace and peak awareness, transporting the meditator to a different layer of reality. It can also be an energizing practice more than a calming one, that can help us rewire our brains to improve our *reception* and *transmission* of intention. I had assumed that intention was like a strong 'oomph', or mental push, through which you project your thoughts to another person to ensure that your wishes are carried out. But the healers described a very different process: intention requires initial focus, but then a type of surrender, a letting go of the self as well as of the outcome.

CHAPTER SIX

In the Mood

MITCH KRUCOFF WAS RETURNING HOME from India in 1994 with almost every idea he had held about the practice of medicine turned on its head. Krucoff, a cardiologist at Duke University Medical Center, and his nurse practitioner, Suzanne Crater, had been invited to inspect the Sri Sathya Sai Institute of Higher Medicine, a hospital in Puttaparthi, at the end of its first year of operation. The hospital was the pet project of the Indian guru Sri Sathya Sai Baba, who wanted to make available the services of a modern Western hospital to the poor and needy, entirely free of charge. Krucoff had been recruited as its cardiac specialist, to advise on the technology needed to build a state-of-the-art facility for high-tech cardiac catheterizations.

Krucoff and Crater were astonished by what they had seen. The overwhelmingly spiritual dimension of the facility – even the special quality of the sound and light – had dwarfed its considerable technological achievements. Spirituality was present in the very design of the building – in the Hindu images lovingly chosen to grace the walls. Situated 9 kilometres from Sai Baba's ashram, the building resembled an elongated Taj Mahal. The wings had been structured as a curvature, like a welcome embrace for all those approaching its doors, and the rotunda inside the entrance was meant to represent a heart whose apex was pointing to heaven.

During their rounds, Krucoff and Crater had been struck by the effect this had on the patients – many of them Indians from extremely remote areas who had never seen running water before. Despite the fact that they had been diagnosed with a life-threatening illness and were set to face an imposing twenty-first-century digital cath lab, not one of them seemed the slightest bit afraid. This utter absence of fear contrasted starkly with the terror and despair to which Krucoff had grown accustomed among the cardiac patients he regularly saw back home.

Krucoff longed to introduce some of these practices to hospitals in America, but if he were going to convince any of his colleagues in cardiology, he would have to prove the benefit of spirituality to the practice of heart surgery through hard data showing a measurable physiological effect. He would have to demonstrate that intangible aspects like intention, or spiritual beliefs, or even a spiritual, uplifting environment, could really make a difference to a patient's outcome.

During the 18-hour flight home, Krucoff and Crater began teasing out ideas for a study. The only way to do it, they eventually realized, was to put prayer to the test – the biggest test of its kind.[1]

When Krucoff got home, he began researching the scientific literature for any evidence that prayer had improved medical outcomes. Fourteen well-conducted trials of prayer had shown a positive effect. In the most famous, published by Randolph Byrd in 1988, a group of born-again Christians outside a hospital had prayed for patients in a coronary care unit. Those who had been prayed for had significantly fewer symptoms, and needed fewer drugs and less medical intervention.[2] A Mid-America Heart Institute study, published around the time Targ published her AIDS study and considered at the time to

have bolstered Targ's findings, showed that Christians of all denominations enlisted to pray for hospitalized cardiac patients reduced
symptoms by 10 per cent, with fewer medical setbacks.[3]

Prayer is viewed as a kind of super-intention, a joint endeavour:
you do the intending, and God carries it out. In some quarters,
intention is considered synonymous with prayer, and prayer synonymous with healing; when you send out an intention, God puts the
intention into action. Indeed, many consciousness investigators
consider these early prayer studies intention experiments. The small
studies that had made use of groups of Christians to send intercessory prayers to heart patients are often construed as a group intention – an attempt by a collection of people to influence the same
thing at the same time.

However promising the results of these early studies, Krucoff
realized that a large-scale trial with tightened protocols was needed,
and he mounted his own small pilot study. He enlisted 150 cardiac
patients, recruited from nearby Durham Veterans Affairs Medical
Center, who had been scheduled for angioplasty and stents. Besides
prayer, Krucoff wanted to see whether 'noetic' therapies, involving
some form of remote or mind-body influence, could affect patient
outcomes. He divided the patient population into five groups. In
addition to standard medical treatment, four of the five were to
receive one of the noetic treatments – stress relaxation, healing
touch, guided imagery or intercessory prayer. The fifth group would
be given no additional intervention besides orthodox medical care.
Every patient would undergo continuous monitoring of brain waves,
heart rate and blood pressure, to gauge the moment-by-moment
effect of these intangible healing influences.

Krucoff decided to turn up the volume on prayer to full blast. To
recruit prayer groups, his nurse-practitioner assistant Suzanne
Crater launched a worldwide campaign of solicitation. She wrote
to Buddhist monasteries in Nepal and France, and to
VirtualJerusalem.com, which arranged for prayers to be placed in the

city's Wailing Wall. She phoned Carmelite nuns in Baltimore to ask for prayers during evening vespers. By the time she finished her campaign, she had enlisted prayer groups from seven denominations, including Fundamentalists, Moravians, Jews, Buddhists, Catholics, Baptists and members of the Unity Church.

Each prayer group was assigned a group of patients, who were identified only by name, age and type of illness. Although Crater and Krucoff left the design of individual prayers to the groups them-selves, they stipulated that the patients had to be prayed for by name and that the prayers on behalf of these patients had to concern their healing and recovery. The prayer portion of the study would be blinded, so that neither patients nor staff knew who was going to be prayed for. The other noetic therapies would be administered an hour after the patients had undergone the angioplasty.

The results were impressive. Patients in all the noetic treatment groups enjoyed 30–50 per cent improvements in health during their hospital stay, with fewer complications and a lower incidence of narrowing of the arteries compared with the controls. They also had a 25–30 per cent reduction in adverse outcomes: death, heart attack, or heart failure, a worsening of the state of their arteries or a need for a repeat angioplasty. But of all the alternative therapies employed, prayer had the most profound effect.

The study was too small to yield any definitive conclusions; after all, only 30 patients had been in the prayer group. Nevertheless, Krucoff's results seemed highly promising. Krucoff and Crater, who had christened their study MANTRA (Monitor and Actualization of Noetic TRAinings), published it and presented their findings before the American Heart Association.[4] Even the most conservative of cardiologists were beginning to take home the message that remote healing might actually work after all, and that prayer in particular was good for the heart.[5]

Krucoff understood that, for his results to be meaningful, the study needed to be replicated on a far larger scale. He rolled out his study and created MANTRA II by launching into an ambitious recruitment programme, eventually enlisting 750 patients from Duke's Medical Center and nine other hospitals across America, and soliciting 12 prayer groups made up of an even larger, more ecumenical collection of the world's major religions. Christians were recruited from Great Britain, Buddhists from Nepal, Muslims from America, Jews from Israel.

Emboldened by his early success, Krucoff and Duke loudly trumpeted the project as the largest multicentre study of remote influence, the supreme test of prayer.

With MANTRA II, he divided the patients into four groups. One group would receive prayer; another, a specially designed programme that included music, imagery and touch (or MIT therapy); the third group, MIT plus prayer; and the final control group, standard medical care. Immediately prior to undergoing angioplasty, those assigned to receive MIT would be instructed in a method of relaxed breathing while visualizing a favourite place and listening to calming music of their choice. They would then receive healing touch for 15 minutes from a trained practitioner. These patients could also wear headphones during surgery.

The point of the new study was to examine whether prayer or the noetic interventions would prevent further cardiovascular events in the hospital, such as death, new heart attacks, a need for additional surgery, readmission to the hospital, and signs of a sharp rise in the enzyme creatine phosphokinase, an indication that the heart has suffered damage. This time, Krucoff also wished to investigate longer-term effects as 'secondary endpoints': whether the interventions could alleviate emotional distress, or prevent death or re-hospitalization at any point six months after the patients had been discharged.

Krucoff's study fell right in the midst of the terrorist attacks of 9/11 and their aftermath. For three months, patient enrolment in the study fell so sharply that he had to amend its design. He developed a 'two-tier' prayer strategy by recruiting 12 'second-tier' prayer groups. As soon as new patients were added to the study, the second-tier groups were to pray for the prayers of the 'first-tier' prayer groups, who had been praying for the patients all along. Through this strategy Krucoff hoped that newly enrolled patients would receive a higher 'dosage' of prayer to approximate the amount received by his patients enlisted earlier in the study.

After the enormous advance publicity, Krucoff's findings were an enormous letdown. When the results were finally in and tallied, there was no denying it: there were no differences in outcomes between any of the various groups during their hospital stay. The only apparent benefit was a slight reduction in distress among the MIT patients prior to the surgery. Otherwise, the large-scale MANTRA was an utter failure. Prayer did not seem to make anybody better.[6]

Among the long-term effects, there had been some therapeutic effects in alleviating emotional distress, need for further hospitalization, and even death rates after six months, but these were not considered statistically significant and they hadn't been the main focus of the study.

Wresting a small victory from this enormous defeat, Krucoff managed to get his findings published in the prestigious British medical journal, *The Lancet*. To the public, he maintained that he was 'thrilled' with the findings and that they had been misinterpreted. Krucoff's study appeared to vindicate the sceptics of prayer as a subject for scientific inquiry. The simple message appeared to be that getting someone to pray for you just does not work.

Meanwhile, in 1997, the Mayo Clinic had begun a two-year study of patients with cardiovascular disease who had been recently discharged from its coronary care unit. Nearly 800 patients were subdivided into two groups: high-risk (those who had one or more risk factors, such as diabetes, a prior heart attack or pre-existing vascular disease) and low-risk (those who had no risk factors other than their present symptoms). The two groups were again divided into two. In addition to ordinary medical treatment, one group in each of the two categories was to receive the prayers of five people once a week for 26 weeks. The two other groups would simply continue with standard medical treatment.

At the end of the study, the investigators concluded that prayer made no difference in mortality, future heart attacks, need for further intervention or hospitalization. Although there were small differences between the treated and untreated groups, particularly among the low-risk patients, these results were not deemed to be significant.[7]

To settle the matter once and for all, Herbert Benson came forward with an ambitious plan. Benson had managed to straddle both mainstream and complementary camps in medicine and was well respected for it – a diplomat with the status of elder statesman between two suspicious factions. Besides his Harvard Medical School credentials, he had set up the Mind/Body Medical Institute, which was devoted to the study and practice of mind–body healing techniques. He had even coined the term 'the relaxation response' to describe their effects.[8] Lending his name to a study of prayer would legitimatize it among the conservative camps.

For this study, Benson recruited five other powerhouses of medicine in the USA, including the Mayo Clinic. His plan was that this study of prayer, which he had dubbed STEP (Study of Therapeutic Effects of Intercessory Prayer), would be the largest, most scientifically rigorous of all time.

The study recruited 1800 patients undergoing coronary artery bypass surgery and divided them into three groups: the first two groups were uncertain whether they were going to receive prayer or not; the first group received prayer and the second did not. The third group, which would definitely receive prayer, was also told of the fact. Benson settled on this particular design so that he could isolate two potential effects: whether being prayed for in itself worked, and whether knowing you were going to be prayed for had any additional benefit. In this way he could control for the effect of belief.[9]

For his prayer groups, Benson enlisted a group of Roman Catholic monks and members of three other Christian denominations: St Paul's Monastery in St Paul, Missouri; the community of Teresian Carmelites in Worcester, Massachusetts, and Silent Unity, a Missouri Unity prayer ministry outside Kansas City. He maintained that his prayer groups included no members of Islam or Judaism because he could not find non-Christian groups happy to work within the demands of the study schedule. The prayer groups were given the patients' first names and the initials of their surnames. Although the design of their prayers could be individual, they had to include the phrase: 'for a successful surgery with a quick, healthy recovery and no complications'. The groups were then followed for 30 days and any post-operative complications, major events or deaths tracked among all groups.

The results shocked the world and bewildered the researchers, most of all Benson, who had spent much of his career promoting the beneficial effects of the mind on the body. The researchers had predicted the greatest benefit in the prayed-for-and-knew-it group, the second greatest effect in the prayed-for-but-didn't-know-it group and the least effect among the didn't-get-prayed-for-and-didn't-know-it group. But their results indicated that no amount of prayer under any condition, whether the patients knew it or not, made any difference to the outcome of their operations. Indeed, the results were the very opposite of the researchers' expectations. Those patients who were prayed for and knew they were being prayed for were worse off, by a statistically significant degree: 59 per cent of the prayed-for-and-knew-it group suffered post-operative complications, compared with 52 per cent among the non-prayed-fors. Even the prayed-for-but-didn't-know-it group suffered slightly more heart attacks and strokes than those who had not been given prayer. Among the uninformed patients who had received prayers, 10 per cent suffered major complications of the surgery, compared with 13 per cent of those who did not receive prayer.[10]

Benson and his co-authors didn't know what to make of these results. They even wondered if the patients had suffered from a type of 'performance anxiety' as a result of the undue pressure and expectations created by the prayers.

Many commentators concluded that this study proved that prayer not only does not work, it is bad for you – or at least it cannot be scientifically tested. Krucoff, who was asked to write a commentary about the study, emphasized that prayer indeed had an effect – a negative one. People needed to discard the universally held view that being prayed for is 'a priori' good for you as these results impelled one to consider that not simply 'voodoo and spells' but also 'well-intentioned, loving, heartfelt healing prayer might inadvertently harm or kill vulnerable patients in certain circumstances'.[11]

The *American Heart Journal* released the study online, and its authors held press conferences. Benson cautioned the media that STEP was not the last word on prayer, although it did raise questions about whether patients should be told about prayers being offered for them. A patient's awareness of being prayed for was considered the most important subject about prayer for future study. But others were not sure whether prayer should or could be studied any more. The John Templeton Foundation had spent $2.4 million on the study, and with negative results like these it was likely that theirs would be the last funds available.

The STEP findings seemed to undercut my own plans for a large intention experiment. Then as I mulled over the negative findings, I came to think that the very designs of the studies might have been responsible. Although the studies attempted to be rigorous, in many instances they violated the most basic rules of scientific research.

For instance, all of the failed studies did not clearly formulate the content of the healing intention, and left the content of the prayers up to the individual supplicant. Although Benson asked that the single phrase 'for a successful surgery with a quick, healthy recovery and no complications' be included, he had not asked them to be specific. The most successful intention experiments incorporate a highly specific target into the intention. In Targ's study, the healers were given the immune system T-cell counts of the AIDS patients and they sent healing specifically to improve the counts. The prayer groups should have been instructed to ask for a specific outcome in cardiac symptoms, or fewer cardiac stents placed during the study time, or any other highly specific request, rather than a nebulous, highly generalized statement about the patient improving.

None of the studies tightly controlled for the number of people involved in the prayer groups or for either the frequency or length of time they were to pray, which again might have confused the mass intention. Perhaps, since they were using highly diverse prayer groups, their prayers were not equivalent. In Benson's study, the prayer groups were allowed to pray anywhere from 30 seconds to several hours four times a week. His researchers never recorded how long the individuals prayed. In Targ's study, although diverse healers were used, they rotated patients, so that each received only a single healing message at any one time.

As Bob Barth, director of the Office of Prayer Research, put it: 'How do you determine a dose of something as intrinsic as prayer? For example, is one 5-minute prayer by a Buddhist different from 10 Catholic nuns in prayer for an hour or more? Is prayer more effective once or 20 times a day?'

In commenting on Krucoff's findings, *The Lancet* also aired its reservations about his study design. 'Could a more restricted denominational approach have influenced the outcome?'[12]

Benson's attempt to standardize the prayer methods used in his study inadvertently interfered with the methods by which the prayer groups usually carry out intercessory prayer. In ordinary circumstances, when prayer groups are asked to pray for someone, they request specific details about the patient, including full name, age, medical condition and periodic reports of the patient's progress. Often they meet with the patient and his or her family. By gathering this personal information, they are able to personalize the prayers.

Benson's study design allowed for the prayer groups only to be given the name and a last initial of the person to be prayed for. The limited information made it impossible for the prayer groups to establish a meaningful connection with or indeed even to zero in on the people they were praying for – one of the conditions that Schlitz and Radin consider important for effective remote influence. Several groups in Benson's study objected to the design of the study. As one commentator wrote, 'This would be similar to the concept of attempting to make a cell phone call to a friend and expecting her to answer when you have only dialled the first three digits of the phone number.'[13]

Like STEP, Krucoff's studies did not reveal anything about the patients in order to create a connection. In Targ's research, the healers had been given a photo and a name as well as information about the patient's condition. None of the groups tested the difference between praying for a patient whose full details were disclosed and simply praying for someone with a first name and last initial.

The selection of the prayer groups was equally unscientific. None of the major prayer studies used any criteria to select participants in the prayer groups or kept track of their size or experience in prayer. Targ had selected only those healers who were highly experienced and committed with a long track record of successfully healing. Although Schlitz's Love Study employed amateurs sending healing intention, training was provided to ensure a homogeneous approach.

Another problem was the lack of a genuine control group in any of the studies. To be truly scientific, a study must be 'randomized' and randomly select participants in one group that is given the treatment and compare its outcome with a group not exposed to the treatment. However, in any health crisis, family members routinely turn to prayer. The odds were overwhelming in all the major prayer studies that the not-prayed-for people were being prayed for by their own loved ones. In MANTRA II, 89 per cent of the patients from both treatment and control groups admitted that someone in their family was praying for them. These patients lived in the religiously active American Bible Belt.

The lack of a pure control group ultimately muddies the results of a study. This problem occurred with the early studies investigating the potential of hormone replacement therapy (HRT) to cause cancer. Many such studies were tainted because it is virtually impossible to enlist women for study who have not taken some form of exogenous hormones – the birth-control pill, the morning after pill or HRT – at some point in their lives. Consequently, none of the studies has a clean control group of true 'non-takers', with which to compare results. Women who take hormones now are compared with women who have taken hormones in the past. Both situations carry a cancer risk. The same 'tainting' would apply to these prayer studies. People in the 'treatment' groups getting prayed for are being compared with patients whose relatives are praying for them.

The large prayer studies had other basic flaws. In both the Benson and Krucoff studies, the people praying did not know the patients and so would not have had a strong motivation to heal, as the 'senders' had in the Love Study. In Benson's study, as Krucoff pointed out in his commentary about STEP, there should have been a true placebo group, which would have no expectation of the possibility of prayer and also there should have been a comparison between such a group and a super-group, whose members included all those exposed to prayer. No analysis compared the effect of being prayed for with the particular belief a patient held about which groups he or she had been assigned, which would have shed light on the possible role of a placebo effect. The researchers also had not taken into account any possible stress on the patient from having to hide his or her assignment in the study from the hospital staff.[14]

Like STEP, Krucoff's study violated the basic rules of scientific design, largely because of events beyond his control. When he reconstituted his study in the wake of 9/11, some of the patients received straightforward prayer from diverse prayer groups, and the others, who had been enrolled after the World Trade Center tragedy, received the 'two-tier' type of prayer, in which those doing the praying were themselves prayed for. Unlike the most basic of scientific trials, his study did not offer the participants the identical treatment.

Even Targ had complained about problems in study design of the very first major prayer study by Randolph Byrd, in which ordinary Christians had been asked to pray for cardiac patients. There was no information about who was taking blood pressure medication, so it was unclear whether prayer or medicine had done the healing. There were no controls for mental attitude during the study. A high number of patients with a positive outlook may have landed in the treatment group. Sometimes a placebo effect, an expectation of healing, can be a large factor in positive results. In one healing study of patients suffering from clinical depression, all the patients improved, even the control group, which did not receive healing, largely from the psychological boost created by the possibility of healing.[15]

In Benson's study the prospect of prayer might have had the opposite effect. According to Larry Dossey, the elegant Southern internist and author of many books on prayer,[16] the STEP study offered prayer as a 'tease', dangled in front of seriously ill patients as something they might or might not be lucky enough to get.

'Nowhere on earth is prayer delivered in this fashion,' says Dossey. 'When prayer occurs in real life, we don't taunt our loved ones with it. They are extended compassionate prayer unconditionally and without equivocation. Who can say what emotions – resentment? hostility? – were generated in these three groups of patients as a result of how prayer was offered?'[17]

The fact that the people who knew they were being prayed for not only had no placebo response but also evidenced more post-surgical complications than any other group, he says, 'suggests that very strange internal dynamics were operating within the Harvard prayer study.'[18]

The Mid-America Heart Institute study – the study in which prayer by Christians of diverse denominations had reduced symptoms in heart patients by 10 per cent – was also criticized for offering so many endpoints that it was bound to show a positive result.[19]

The negative results of these large prayer studies could be because praying for others does not work, because prayer simply cannot be subjected to scientific study, or simply because these new studies themselves were asking the wrong questions. After all, according to Bob Barth of the Office of Prayer Research, these studies only represent a small proportion of prayer research.[20] Of the more than 227 studies investigated by the office, 75 per cent show a positive impact.

Nevertheless, to study the effect of remote intention, it may be best to move away from prayer, which contains a good deal of emotional baggage. Targ tried to isolate the effect of simple healing intention, which is different from prayer. With intention, the agent of change is human; with prayer it is God. Simple healing intention can be more easily controlled for in a scientific study by ensuring that every member of the group sending the intention was sending the exact same message. For the purposes of my intention experiments, a simple intention to heal or improve something might avoid all the problems associated with studying prayer. Unlike prayer, healing has been persuasively proven; a large body of evidence exists about the positive effects of distant healing – perhaps 150 studies in all.[21] These scientific studies have been subjected to overall reviews that rate both the significance of the effects and the outcome. In the most cautious of such analysis, Professor Edzard Ernst, the exacting and sceptical chair of complementary medicine at Exeter University in Britain, concluded that of 23 studies, 57 per cent had shown a positive effect.[22] Among the most rigorously scientific (those with double-blind trials), the average effect size, or size of change among those treated, was 0.40 – about 10 times better than the effect size of aspirin or propanolol, two drugs considered highly successful in preventing heart attacks.

Hidden in the failure of the large prayer studies lies vital instruction not only about the design of such mass experiments, but also about those elements that maximize the power of intention. To be successful, an intention may require other parameters besides trained attention, getting out of the way, and formulating a simple request to the universe. As Gary Schwartz learned during his own research on healing, the attitude of the healers as well as the patients may matter a good deal.

Schwartz's research began as a simple study of healing intention by Reiki practitioners. Schwartz had enlisted his colleague, Beverly Rubik, founding director of the Center for Frontier Sciences at Temple University, Philadelphia, a biophysicist interested in subtle energies. As Rubik was well versed in studies using bacteria, they decided to use as their subject *E. coli* bacteria, which had been severely stressed. One way to stress bacteria is to shock them with a sudden blast of heat. Schwartz, Rubik and their colleague Audrey Brooks carefully managed the amount of heat so that it was enough to stress the bacteria without killing off the entire sample. They then asked 14 practitioners of Reiki to heal the bacteria that survived by transmitting a standard Reiki treatment for 15 minutes. Each practitioner was to heal three different samples over three days. Equipment with an automated colony counter kept track of the number of bacteria that survived.

Initially, Schwartz, Rubik and Brooks were surprised to find that the Reiki practitioners made no difference to the overall survival of the viable bacteria. On closer look, however, they discovered that the Reiki practitioners seemed to be successful on certain days, but not on others. This spotty batting average puzzled them. Perhaps, Schwartz thought, a healer's success depended on some sort of connection with the subject. It was difficult, after all, to feel any warm and fuzzy connection with *E. coli* bacteria, which ordinarily resides peacefully in the gut but can wreak havoc when it migrates out of the digestive tract. But what if he managed to get his practitioners in healing mode?

In the next batch of studies, Schwartz and his colleagues asked the Reiki practitioners to work for 30 minutes on a human patient suffering with pain, and then set them back to work on their bacteria samples. This time, the healing was successful; the scientists discovered significantly more bacteria in the healed samples than in the controls. The healers appeared to enjoy a higher success rate once their healing 'pumps' had been primed.[23]

Nevertheless, Schwartz and the other researchers continued to discover instances in which the healers had a deleterious effect on the bacteria. It occurred to them that a healer's own well-being might affect results. They needed a simple test to assess true well-being, to gauge more than physical condition. They decided to use the Arizona Integrative Outcomes Scale (AIOS), an ingeniously simple visual means of assessing spiritual, social, mental, emotional and physical well-being during the past 24 hours.[24] Developed by physician and psychologist Iris Bell, one of Schwartz's colleagues at the University of Arizona, AIOS allows patients to assess more than physical symptoms. The subjects are told to reflect on their general sense of well-being, 'taking into account your physical, mental, emotional, social, and spiritual condition over the past 24 hours', then to mark a point on a horizontal line between 'worst you have ever been' on the left and 'best you have ever been' on the right that, in their view, represents their overall sense of well-being in the same time period. A number of studies demonstrated that AIOS is a useful, accurate tool for pinpointing emotional wellness and a healthy state of mind.[25]

In their next series of studies, Schwartz, Rubik and Brooks asked the Reiki healers to assess themselves on the AIOS scale before and after they had carried out the Reiki. With this data, the scientists discovered an important trend. On days when the healers felt really well in themselves, they had a beneficial effect on the bacteria; the counts in the bacteria given the therapy were higher than in the heat-shocked controls. On days when they did not feel so well and they scored lower on the test, they actually had a deleterious effect. Those practitioners who began the healing with diminished well-being actually killed off more bacteria than naturally died in the controls. Evidently, a practitioner's own overall health was an essential factor in his ability to heal.

Schwartz and his colleagues then tried a study using AIOS with a different type of healing, called Johrei. They recruited 236 practitioners and volunteers, and asked them to fill in the AIOS scale plus a questionnaire he had created assessing emotional state of mind before and after they administered healing. When Schwartz and Brooks compared the AIOS tests of both the healers and the patients before and after the healing, they discovered another interesting effect. Although the patients felt better after they had received the healing, so did the healers after they had performed the healing.

Giving was as good as getting for these senders. Other research showed a similar result.[26] The act of healing and perhaps the healing context was itself healing. *Healing someone else also healed the healer.*[27]

Schwartz and his fellow researchers then carried out another study of distant Johrei healing on cardiac patients – a double-blind study so that no one but the statistician knew who was receiving healing.[28] The primary outcomes measured were clinical reports of pain, anxiety, depression and overall well-being. After three days, the patients were asked if they had had a sense, feeling or belief that they had received Johrei healing. In both the treatment and control groups, certain patients strongly believed that they had received the treatment and others had a strong feeling they had been excluded.

When Schwartz and Brooks tabulated the results, a fascinating picture emerged. The best outcomes were among those who had received Johrei and believed they had received it. The worse outcomes were those who had not received Johrei and were convinced they had not had it. The other two groups – those who had received it but did not believe it and those who had not received it but believed they had – fell somewhere in the middle.

This result tended to contradict the idea that a positive outcome is entirely down to a placebo response; those who wrongly believed they received the healing did not do as well as those who rightly believed they had received it.

Schwartz's studies uncovered something fundamental about healing: both the energy and intention of the healing itself and the patient's belief that he or she had received healing promoted the actual healing. Belief in the efficacy of the particular healing treatment was undoubtedly another factor. In the Love Study, Schlitz and Stone had stressed the importance of a shared belief system in the success of remote influence, and Schwartz's results bear this out.

In the large prayer studies, the senders and receivers of prayer did not share the same belief system about God. Most of the patients had been prayed for by a number of groups from different religions and disparate belief systems. Even Benson's Christian study employed different Christian sects, which do not share identical beliefs. It may be uncomfortable for some groups to be prayed for by people who do not share their views about the divine.

As Marilyn Schlitz pointed out, none of the clinical trials made use of what scientists call 'ecological validity'. This means that the trials were not designed to model what happens in real life. In the Harvard study, for example, the prayer groups were instructed to pray differently from how they would normally. None of the big prayer studies tested the effect of the kind of prayers that prayer groups believe is most likely to work.[29] In these studies, says Dossey, 'what is being tested is not genuine prayer but a watered-down faux version of it'.[30] The contents and context of prayer were treated casually, as if it were no different than some new medication. The Benson study also framed its intention as a 'negative' – asking that the patients heal with 'no complications' – countering the most basic folklore about prayer and affirmations, which stipulates that they should always be framed as a positive statement.

Ordinarily, says Schiltz, people have a meaningful relationship with the person they are praying for. Psychologist and mind-body researcher Jeanne Achterberg, of the Institute for Transpersonal Psychology in California, carried out a study at a Hawaiian hospital, using highly experienced distant healers, who selected as their 'patient' a person with whom they had a special connection. Each healer was isolated from his patient, who was then placed in an MRI scanner. At random, two-minute intervals, the healers sent healing intentions to their patients, using their own traditional healing practices. Achterberg discovered significant brain activation in the same portions of the brains – mainly in the frontal lobes – of all the patients during times healing energy was being 'sent'. When the same regime was tried out on people the healers did not know, they had no effect on the patients' brain activity. Some sort of emotional bond or empathetic connection may be crucial to the success of both prayer and healing intention.[31]

The large prayer studies may have failed because the researchers were looking in the wrong places for demonstration of an effect. A study of AIDS about to be published at the time of writing has also failed to find an effect. Nevertheless, a highly significant number of people in the treatment group correctly guessed which group they were in, while the control group did not. As Schlitz concluded, 'The treatment group seemed to feel something; it just did not correlate with the clinical outcomes that were measured.'[32] The study may just have been asking the wrong questions.

Another important variable may be the kinds of thoughts experienced by the recipient during healing. Researchers have discovered that negative thoughts and visualization can have a powerfully negative effect on the body, as if the negativity is somehow infectious and these thoughts take physical form. For instance, Pennsylvania researchers from the Center for Advanced Wound Care in Reading, Pennsylvania, have discovered that patients with slow-healing wounds often have negative thought patterns and behavioural or emotional wounds, such as guilt, anger and lack of self-worth.[33]

The same effect can occur with negative relationships. A recent study of couples showed that the stress of reliving an argument delays wound healing by at least a day. In an ingenious study by Ohio State University College of Medicine, the researchers gathered together 42 married couples and inflicted small wounds with a tiny puncture device on one partner of each pair. During the first sessions, the partners held a conflict-free, constructive discussion and the wound healing was carefully timed. Several months later, the researchers repeated the injury, but this time allowed the partners to raise an ongoing contentious issue, such as money or in-laws. This time, the wounds took a day longer to heal. What is more, among the more hostile couples, the wounds healed at only 60 per cent the rate of the more compatible pairs. Examination of the fluids in the wounds found different levels of a chemical called interleukin-6 (IL-6), a cytokine and key chemical in the immune system. Among the hostile couples, the levels of interleukin-6 were too low initially and then too high immediately after an argument, suggesting that their immune systems had been overwhelmed.[34]

The person sending out an intention might also need to be sent good intentions. Krucoff's results as universally interpreted had overlooked one vital finding: the patients with the double-tier prayer groups who had been prayed for had fared far better in the secondary endpoints; their death and re-hospitalization rates over the six months after discharge were 30 per cent lower than the others. Mortality over six months was lower among patients given MIT, and lowest of all among patients given MIT with prayer. These results had only been characterized as a 'suggestive trend', but may have been the entire point of the story. *Praying worked if the person doing the praying – or his prayers – also had been prayed for.*[35]

Healing and positive intention are simply an aspect of the constant two-way flow of communication between living things. In the person being sent intention, a shared belief in the power of the healing modality and a positive state of mind may enhance results. Fritz Popp's research demonstrates that the degree of coherence of an organism's light emissions is linked to its overall state of health. When healers are healthy, in a positive state of mind and have engaged in a healing 'warm up', their light is more likely to shine brighter. The most effective healer of all may be the one who has been healed himself.

CHAPTER SEVEN

The Right Time

AROUND LAURENTIAN UNIVERSITY CAMPUS in Canada, Michael Persinger's basement vault was known as the Chamber of Heaven and Hell. Room COO2B, a disused sound booth, was a relic of the 1970s, its original fittings intact: enormous nylon loudspeakers, deep orange flecked shag carpeting and a single item of furniture – a stained brown polyester armchair. More than 2000 people had occupied the chair in pure darkness, a modified yellow motorcycle helmet on their heads, surrendering all control of their next half hour to the scientists behind the glass booth. Persinger, a neuroscientist, was god of room COO2B. He had become expert in manipulating brain waves to yield up a divine experience, or, as he referred to it, 'a sensed presence'. With a few simple commands typed into a computer, he would instruct the helmet to send low-level magnetic fields coursing through the temporal lobes of his volunteers, abruptly switching sides of the brain to heighten the transcendent and occasionally terrifying nature of the experience.[1]

Jesus had been sighted in the brown polyester reclining chair, as had the Virgin Mary, Muhammad, monks in hooded robes, knights in shining armour and a Native American deity, the Sky Spirit. Out-of-body experiences had been produced; near-death experiences relived. One journalist had been transported back to his life's most transcendent moment – the time he first laid eyes on his high-school girlfriend's perfect breasts.

Not all visitors found God. There had been imaginings of alien sightings and abductions, and even satanic ritual. One volunteer, overwhelmed by the sight of an enormous set of eyes and the smell of burning sulphur, attempted to pull himself loose from the helmet and wrench off the blindfold and earplugs. As soon as the 500-pound door was pried open for him he fled, terrorized, from the room.

The nature of the experience all depended, Persinger and his assistants explained, on a physiological roll of the dice: the sensitivity of the left amygdala of the brain compared with its counterpart on the right. If the left is more sensitive, and you send magnetic waves coursing through it, you get heaven. If you are unlucky enough to be born with a more sensitive right amygdala, you get hell.[2]

Persinger had a singular passion: the subtle influences of geology and meteorology on human biology, particularly the electrical circuitry of the brain. A transplant from the American South, he had headed north in the 1960s to avoid the draft and a likely stint in Vietnam – a possibility he objected to on moral grounds – and he remained in Canada after receiving a professorship at Laurentian in 1971. Forty years later, he seemed an unlikely draft dodger, with his three-piece pinstripe suits, gold-chain swag and watch fob, and clipped, offhand manner. This conservative posturing masked a bold curiosity that led him into exotic areas of inquiry – the rhythms of biological systems, the volatile energy of outer space, the nature of epilepsy, the source of mystical visions – disparate areas that eventually converged in his mind after an extraordinary epiphany. Persinger realized that living things are attuned not only to each other, but also to the earth and its constantly shifting magnetic energies. This remarkable revelation, built upon the discoveries of Franz Halberg, would convince me that careful timing to coincide with these energies might be vital for an effective intention.

In 1948, as a young medic at Harvard Medical School on a temporary visa from war-torn Austria, Franz Halberg was assigned an impossible task: to help find the cure for all disease.[3] At the time, the cure was assumed to involve the cortical hormones secreted by the adrenal glands, which enable the body to adapt to the ordinary stresses of life. The search was on to find reasonable substitutes for the body's own scarce supply of steroids.

Halberg had been singled out to study mice whose adrenal glands had been removed and who were then injected with adrenaline in order to observe the effect on their circulating white blood cells called eosinophils. In ordinary circumstances, adrenaline will set off a predictable seesaw, causing more of the body's natural steroids to be secreted, which, in turn, lower the eosinophil count. However, in animals or humans without adrenal glands, the count should remain static. But the cell count in Halberg's mice still seemed to fluctuate, even after he had removed all trace of adrenal tissue. Later, after moving to the University of Minnesota, he carried on his studies with a near limitless supply of experimental mice, and came up with the same conclusions. Even when he handled them less frequently, which should have caused less stress to the tiny creatures, he noticed more variation in cell count.

Halberg was mystified by this fluctuation, until he suddenly recognized a recurring pattern: the cell counts were always higher in the morning and lower at night. The variation was rising and falling according to a predictable, 24-hour cycle. Halberg studied other biological processes, and discovered that many appear to run according to an in-built clock. All living things respond to the same 24-hour rhythm, in tandem with the earth's rotation. Halberg coined the terms 'chronobiology' – the influence of time and certain periodic cycles on biological function – and 'circadian' (circa = about; dia = day) for daily biological rhythms. He created the Chronobiology Laboratories at the University of Minnesota and became known as the father of chronobiology. Chronobiology, as his lab began to discover, is a ready-made feature of organisms, not simply something learned or acquired – an inherent property of life.

Besides circadian rhythms, Halberg also discovered that living things keep in time to many other periodic rhythms; half-weekly, weekly, monthly and yearly cycles govern virtually every biological function. The human pulse and blood pressure, body temperature and blood clotting, circulation of lymphocytes, hormonal cycles and other functions of the human body all appear to ebb and flow according to some basic, recurring timetable. These rhythms are not unique to humans, but are present throughout nature, and evident even in fossils of single-cell organisms that had existed millions of years ago.

Initially Halberg believed that the master switch for these biological rhythms was located in certain cells of the brain or adrenal glands. However, certain cycles carried on even when Halberg removed the brain cells in question – the adrenal glands – and even the brain itself. In his eighties, Halberg made his final breakthrough discovery: the synchronizer within every living thing is not internal but resides in the planets and in the sun.[4]

The sun is a furious star. This huge ball of gases, with a surface temperature of around 6000°C, is encased by strong magnetic fields in the outer solar atmosphere – a recipe for periodic explosions, as the gases build up and magnetic fields intersect on the sun's surface. Although the patch of space between sun and earth used to be considered an uneventful vacuum, 'space weather' is now understood to be weather so extreme, of such unimaginable turbulence, that if transferred to earth it would blow up the entire planet in an instant.

Solar wind, a constant blast of electrified gas, dominates this inter-planetary medium, soaring past the earth at speeds up to 2 million miles per hour. Although the earth's magnetic field usually deflects it, this gale can penetrate our magnetic field during moments of intense solar activity.

Sunspots – vortices of concentrated magnetic fields, visible to us as dark blobs on the sun's surface – begin to accumulate and then to disappear in fairly regular cycles, so that scientists can make some predictions about when the sun is likely to erupt. A solar cycle of waxing and waning activity occurs, on average, every eleven years. As sunspots build up, so does the sun's aggressive behaviour. At unpre-dictable moments, it hurls solar flares, gaseous explosions with the energy of 40 billion atomic bombs, likely caused by the ripping apart and reconnection of strong magnetic fields. Electrified bullets of high–energy protons from the nuclei of gases are picked up by the solar wind and flung towards earth at speeds of more than 5 million miles per hour, showering our atmosphere with radiation and ioniza-tion. Periodically, the sun also releases a corona mass ejection, a ball of gas and magnetic fields of up to a billion tons, which also speed towards earth at several million miles per hour, causing extreme geomagnetic storms in space.

Scientists have long understood that earth is, in effect, a giant magnet with two poles – North and South – surrounded by a magnetic field that is constantly in flux. This field encircles the earth like a donut in a region of space called the 'magnetosphere', and is kept in place by the solar wind, with a force of about 0.5 gauss or 50,000 nanotesla – about 1000 times weaker than that of a typical horseshoe magnet.

The geomagnetic fields (GMFs) differ in different regions and at varying times. Any changes in our solar system – the activity of the sun, the movement of the planets, the daily oscillation of the earth on its rotation – or geological changes on earth – the presence of ground water or the movement of the earth's molten inner core – can alter the strength of the earth's GMF on a daily basis. Storms in space transfer some of the energy of the solar wind to the earth's magnetosphere, causing wild fluctuations of direction and speed in the particles in the earth's magnetic field. The National Oceanic and Atmospheric Administration (NOAA), which tracks these volatile space weather patterns, reckons that over any given solar cycle, geomagnetic storms in space will occur about a third of the time, almost half of which are severe enough to interfere with modern technology. Storms of this magnitude (G5, or maximum severity on the NOAA scale) can disrupt portions of the earth's electrical power, pipeline flow and high-tech communications systems, and disorient spacecraft and satellite navigation systems. In March 1989, one such storm left 6 million people in Montreal without electric power for nine hours.

At the time Halberg made his discoveries, geomagnetic storms were known to have a profound effect on the movement and orientation of animals such as pigeons and dolphins, which make use of the earth's geomagnetic field to navigate. Biologists assumed that the earth's weak magnetic field had little effect on basic biological processes, particularly as living things have daily exposure to the more powerful electromagnetic and magnetic fields generated by modern technology. But in the course of investigating the health implications of space flight, the Soviet researchers uncovered evidence that natural geomagnetic fields, particularly those of extremely low frequencies (less than 100 hertz), have a pronounced effect on virtually all cellular and chemical processes in living things.

When Russian scientists at the Space Research Institute of the Russian Academy of Sciences explored the effects of space weather on cosmonauts being sent into space, they discovered that protein synthesis in bacteria cells is highly susceptible to changes in geomagnetic fields, and that this disturbance in protein synthesis also affects human micro-organisms.[5] Geomagnetic disturbances influence the synthesis of micronutrients in plants; even single-celled algae respond to solar-cycle flux.[6] So attuned are plants and micro-organisms to these changes that the Russian researchers made use of them as a sensitive barometer for geomagnetic disturbances.[7]

The Soviet scientists also discovered that if the cosmonauts suffered cardiac arrest, it was usually during a magnetic storm.[8] Illness on earth also appeared to parallel geomagnetic activity in space; both sickness and death increased on stormy geomagnetic days.[9] But of all the systems in the body affected, changes in solar geomagnetic conditions most disturbed the rhythms of the heart.

The Space Research Institute scientists tracked the heart rate of healthy volunteers over an entire solar cycle and compared it with sunspot and other geomagnetic activity during that period. The healthiest heart rate is one with the greatest variation. In the Russian research, the most varied heart rate occurred during times of the least amount of solar activity,[10] while heart rate variability (HRV) decreased during magnetic storms. A disturbance in HRV most affects the autonomic nervous system, the system in the body that keeps it ticking over without any conscious intervention. A low HRV increases the risk of all coronary artery disease and heart attack. During increased geomagnetic activity, the viscosity, or thickness, of the blood also increases sharply, sometimes doubling, and the blood-stream slows down.[11]

Sudden cardiovascular death also appears to be linked with solar geomagnetic activity.[12] Heart-attack rates rise and fall according to solar-cycle activity:[13] the largest number of sudden deaths from heart disease occurred within a day of a geomagnetic storm.[14] Halberg himself discovered a 5 per cent increase in heart attacks in Minnesota during times of peak maximum solar activity.[15]

It is not surprising that biological systems like human beings are sensitive to external signals, such as geomagnetic disturbances. Magnetic fields are caused by the flow of electrons and atoms with charge, known as ions, and whenever magnetic forces change, they alter the direction of the flow of these atoms and particles.

Ultimately, since living organisms are also composed of particles like electrons, any profound change of magnetic direction may markedly alter their biological processes.

Once Halberg understood the effect of the earth's geomagnetic field on living things, he renamed his life's work 'chronoastrobiology' – the rhythms of biology as affected by astral bodies. The sun was the giant metronome setting the pace for all of life.

Persinger's interests had mostly to do with geomagnetic effects on the brain. Researchers in the Soviet bloc had also discovered that space weather can affect neurological processes. Scientists at the Azerbaijan National Academy of Sciences at Baku used a special device enabling them to continuously monitor the electrical activity of the heart and brain in a small number of healthy volunteers, and to compare those rhythms with those of the earth's geomagnetic field.

They discovered that geomagnetic activity has a strong influence on brain functioning. During magnetically stormy days, EEG readings get destabilized.[16] Geomagnetic turbulence also disturbs the balance between certain parts of the brain and profoundly disrupts communication within the nervous system, over-activating certain aspects of the autonomic nervous system and lowering others.[17]

The sun's activity also affects mental equilibrium. As Persinger discovered, the more unsettled the weather in space, the greater the number of patients hospitalized for nervous disorders and the greater number of attempted suicides.[18] Geomagnetic disturbance also seemed to correlate with increases in general psychiatric disorders.[19] Even those already suffering from mental illness get more agitated during magnetically stormy days.

Persinger grew intrigued by a possible relationship between geomagnetic fluctuations in the earth and the timing of epileptic seizures after his neuroscientist colleague Todd Murphy, who had temporal-lobe epilepsy as a young child, disclosed that he often had out-of-body experiences while having a seizure. Some data had already linked an increase in geomagnetic activity with the timing of epileptic seizures.[20] Could an epileptic fit result from geomagnetic disturbance? Persinger decided to study this possibility in an animal. He injected a batch of laboratory rats with lithium pilocarpine, which causes epileptic-like seizures in the rodents, and compared the timing of the onset of seizures about an hour after the onset of laboratory-simulated increased geomagnetic activity.[21] From this, Persinger inferred that, above a certain threshold of geomagnetic activity, epilepsy is more likely to be triggered. Whenever geomagnetic activity exceeded 20 nanotesla, seizures would occur more frequently.[22]

Persinger then discovered a relationship between sudden death – from epilepsy or cot death – and high levels of geomagnetic activity.[23] Sudden, seemingly inexplicable deaths might have a rational explanation after all: people with weaker constitutions are at the mercy of the sun's restless activity.

Strong geomagnetic fields also appear to affect learning profoundly – often for the better. Increased geomagnetic activity enhances memory: rats exposed to geomagnetic fields learn mazes more quickly.[24] Large fluctuations in solar activity cause other subtle effects in human behaviour and performance – for instance, the ability to perform a skilled task.[25] Psychologist Dean Radin once examined the effect of GMFs on bowling. He tracked the performance of experienced bowlers over a number of periods, and then compared their scores with the geomagnetic activity of the same period. Large geomagnetic fluctuations the day before a match appeared to cause more uneven results than normal – a 41 per cent variance in the men's scores compared with the more consistent scores obtained during days of geomagnetic stability.[26] Other research has demonstrated that the greater the change in the earth's geomagnetic field, the greater the number of traffic violations and industrial accidents.[27] The most important determinant appeared to be large *change* in geomagnetic activity, either from turbulent to calm or the reverse.

Although periodically destabilizing, exposure to the daily ebb and flow of earth's geomagnetic activity may be essential to life here. The Solar Terrestrial Influences Laboratory at the Bulgarian Academy of Sciences in Sofia carried out biological experiments on board the Soviet Mir space station to examine what happens to cosmonauts who are deprived of contact with the earth's geomagnetic field while in space. The scientists constructed a 'geomagnetic vacuum', a six-metre stainless steel decompression press-chamber, which partially blocked out the earth's natural geomagnetic field. Seven healthy young men were sealed off in the chamber and their bodily processes analysed. After being placed in the decompression chamber, the men evidenced a number of upsets in brain-wave activity. Sleep was more restless, with fewer periods of deep sleep.[28]

Contact with geomagnetic fields may play a primary role in maintaining the equilibrium of the nervous system. Indeed, the earth's tiny geomagnetic fluctuations have the most profound effect on the two major engines of the body: the heart and the brain.

Persinger went on to discover other extraordinary geophysical effects on human beings. Electromagnetic and geomagnetic phenomena resulting from the earth's shifting plates, earthquakes, or from unusually high rainfall levels – even electromagnetic 'luminosities', or lights in the sky – can all stimulate certain portions of the brain that produce hallucinations. Between 1968 and 1971, more than 100,000 people reported observing visions of an apparition of the Virgin Mary above a church in Zeitoun, Egypt. When Persinger examined the seismic activity in the area over the same time period, he discovered an unprecedented peak in earthquake activity.[29] Sometimes the electromagnetic effects were man-made. At one point he studied a Roman Catholic woman with early brain trauma who reported nightly visitations by the Holy Spirit. Ultimately, he discovered the source of the miracle: her disability caused her to be unduly affected by the electric alarm clock situated near her head as she slept.[30]

Persinger wondered whether he could reproduce these types of geomagnetic disturbances in the laboratory. His colleague Stan Koren modified and wired up a motorcycle helmet (thereafter named the 'Koren' helmet) so that it could send out very-low-frequency complex magnetic fields – about the amount that radiates from a telephone handset – in precise directions. Participants would be fitted in the helmet, then placed in the acoustic chamber of room COO2B, which had been especially adapted to block out electromagnetic noise. Turning on the helmet would produce what Persinger referred to as 'temporal lobe transients', something possibly like a micro-seizure – tiny episodes causing alterations in neuronal firing patterns. This produced virtually the same effect on the brain as exposure to increased ambient geomagnetic activity.

Over time, Persinger began to recognize patterns. The brain waves of his participants would fall into resonance with the complex magnetic fields and remain in synchrony for up to 10 seconds after he had turned off the helmet.[31] Through trial and error, he discovered that the portion of the brain most susceptible to electromagnetic and geomagnetic effects are the temporal lobes. Sending low level (1 microtesla), pulsed magnetic fields over the right cerebral hemisphere slowed brain waves to an alpha rhythm (8–13 hertz), but only on the right side.[32]

Our 'sense of self' and our sense of the 'other' are housed in both temporal lobes but primarily in the left hemisphere, where the language centres are located. To function normally, both left and right temporal lobes must work in harmony. If something upsets this balance, the brain will sense another 'self' and create a hallucination. As Persinger discovered in his experiments, stimulating the right temporal lobe portion of the brain generates the sense, presence or feeling of spiritual visions, both good and bad. Aiming magnetic fields at the amygdala of the brain at the same time colours the experience with intense emotion, just as occurs during a spiritual experience. By first stimulating one side of the amygdala and then the other, Persinger found that he could heighten the emotional complexion of the experience.

Volunteers wearing the Koren helmet experienced divine epiphanies, apparitions, out-of-body sensations and even a hallucination of Satan purely through temporal-lobe stimulation. The nature of the experience largely depended on the participant's individual history: negative early life experiences tend to increase the sensitivity of the right temporal lobe, and those with a high proportion of such experiences tend to have a negative experience while wearing the helmet. A happier person, with a more sensitive left temporal lobe, is more likely to experience a sense of the divine.[33]

It would have been tempting for Persinger to conclude that all spiritual experience is simply geomagnetically induced hallucination, except for one unsettling fact: extrasensory perception and other psychic abilities appear to be more acute during particular types of geomagnetic activity. When the earth is 'calm' and geomagnetic flux at an ebb, telepathic and extrasensory perceptions increase.[34] Even minor environmental changes – from slight variations in the weather to solar patterns – appear to have a profound effect on extrasensory perception or the ability to view things remotely. The reverse occurs with psychokinesis – mental attempts to change physical matter. The power of intention increases when the earth's energy is agitated.[35]

In the 1970s, Persinger was able to test the effects of geomagnetic activity on telepathy during sleep by teaming up with noted parapsychologist Stanley Krippner, then the director of a dream laboratory at Maimonides Medical Center in New York City. Krippner had perfected an experimental protocol to test telepathy, clairvoyance and precognition in dreams during deep sleep. Volunteers would be paired off. While one partner slept, the other would be in a separate room and would be asked to concentrate on an image and attempt to 'transmit' the image to the dreamer, so that it would be incorporated into his dream. Upon waking, the participants who had been sleeping would describe their dreams in great detail, to determine whether they contained anything resembling the target pictures they had been sent during their slumbers.[36]

Persinger and Krippner found that participants did better on certain days than on others. When they tracked geomagnetic activity during the period of the study, they discovered that the dreamers had significantly higher accuracy in picking up the target pictures on nights when the earth's GMF activity was relatively quiet.[37]

Geomagnetic activity also affects precognitive dreams – those that forecast events. Dr Alan Vaughan, a well-known clairvoyant whose dreams accurately foretold the future in great detail, kept a detailed dream diary in order to compare their contents with future events. One of Vaughan's dreams predicted the murder of then-presidential candidate Robert Kennedy two days before he was assassinated.[38] An examination of the geomagnetic activity on the nights that Vaughan had dreamed 61 such premonitions showed that it was significantly quieter on the days when he had his most accurate dreams.[39]

During days of geomagnetic calm, spontaneous instances of telepathy or clairvoyance are more likely to occur[40] and remote viewing accuracy appears to improve.[41] Persinger carried out his own intriguing test of ESP using a group of couples. One member of each pair was shown an image while it was being bathed in magnetic fields, then asked to describe the memory of an experience he or she had shared with the partner that was prompted by the image. Simultaneously, in another room, the partner was shown the same images and also asked to describe a memory. When Persinger compared the results, he discovered that the two narratives were most alike when the ambient geomagnetic activity was at its quietest. The greater the geomagnetic activity, the less the two sets of memories mirrored each other.[42]

Nevertheless, the two sexes appear to respond very differently to geomagnetic activity, which Persinger discovered after comparing a database of paranormal experiences with geomagnetic activity and breaking down the data by sex. Men tended to have more premonitions on days when geomagnetic activity was high (above 20 nanotesla), whereas women reported more premonitions if the geomagnetic activity was low (below 20 nanotesla). Men also tended to have more accurate memories with higher geomagnetic activity; women, with lower geomagnetic activity. Just as Krippner had found, the people most susceptible to extrasensory experiences were those with 'thin boundaries', particularly those who had already had paranormal encounters.[43]

With time, Persinger found that he could enhance powers of extrasensory perception with the artificial geomagnetic fields of the Koren helmet. The remote-viewing ability of one of his students considerably improved after he was exposed to weak horizontal magnetic fields.[44]

In 1998, Persinger decided to put the Koren helmet to the ultimate test. Could it interrupt the ability of one of the greatest remote viewers in the world? He invited Ingo Swann to his basement lab. Swann, then 68, soon proved he had lost none of his extrasensory prowess; he correctly described and drew in great detail images of randomly selected photographs sealed in envelopes in another room. Nevertheless, after Persinger bathed the photos in complex magnetic field patterns, Swann's accuracy suddenly plummeted. The most disruptive fields had different signal wave forms of varying phases. This suggested that Swann was picking up the information in wave form and that those signals were easily interrupted by magnetic fields that could disturb their coherence.[45] As Gary Schwartz had also discovered, information transmitted or received by human beings must have a strong magnetic component.

Persinger's evidence persuaded me that geomagnetic activity influences the clarity of our reception in picking up quantum information. But do geomagnetic fields also affect the strength of our transmissions and their effect on the physical world? Research by Stanley Krippner offers a few clues. Krippner wished to test the hypothesis that psychokinesis is likely to occur on days when the earth is 'noisy'. He and his team worked with the Brazilian sensitive Amyr Amiden, known for his extraordinary psychokinetic ability, and set about comparing the time of Amiden's psychokinetic activities with geomagnetic fluctuations in the Brasilia area, where the sessions were taking place. Krippner's team also took readings of Amiden's pulse and blood pressure.

The team found a significant correlation between Amiden's psychic feats and the daily geomagnetic index for the entire southern hemisphere. For instance, Amiden performed the highest number of psychokinetic feats on 10 March and 15 March, which were the days that month with the greatest geomagnetic activity. He produced nothing out of the ordinary on 20 March, the geomagnetically quietest day of the month.[46]

Amiden's psychic abilities were preceded by both a rise in his diastolic blood pressure (the pressure of the blood as it returns to the heart) and a rise in geomagnetic 'noise'. It may be that geomagnetic activity must first cause changes in the 'heart brain' before a person can transmit information that can affect physical matter.

Interestingly, as with couples in the Love Study, Amiden's most powerful psychokinetic effects *anticipated* strong input: in his case, geomagnetic flux. In one instance, two religious medallions suddenly materialized in the room where Amiden and the researchers were present, appearing to drop from the ceiling – an event that was followed by a sudden rise in the area's geomagnetic field. Can humans anticipate this geomagnetic noise, and, if so, do such anticipatory windows offer them more psychokinetic power than usual?

Psychologist William Braud carried out some intriguing studies of the effect of geomagnetic fields on intention by examining whether high levels of geomagnetic activity were correlated with powers of remote influence. Braud examined the effect of sending intention to human blood cells and to another person. Like Krippner, he discovered that the success of intention was linked to a 'noisy' sun producing high geomagnetic activity.[47]

Besides solar activity, other environmental factors should be considered when working out the best times to send intention. A number of scientists, including Persinger, found that certain days and certain times of day influence the success of ESP and psychokinesis.[48] The best results occur around 1 p.m. local sidereal time, which is time measured by our relation to the stars, not the sun. Local sidereal time is worked out as the hour's angle of the vernal equinox, where the plane of the earth's equator would intersect with that of its orbit, if measured out in the heavens. Psychokinetic effects also seem to be greater about every 13 days, at times when solar wind is modulated.[49]

It might also be worth avoiding times of low visibility and high winds, a condition which produces a high percentage of ions with electrical charges in the air. An ion forms when a molecule encounters enough energy to unleash an electron. They are also created by rainfall, air pressure, forces emitted during a waterfall and the friction from large volumes of air moving rapidly over a land mass, as during so-called ill winds, such as El Niño or Santa Anas of southern California. Both positive and negative ions are equivalent to a tiny pulse of static electricity, and the air that we breathe is made up of billions of these tiny charges.

Good 'clean' air contains 1500–4000 ions per cubic centimetre, and the preferred ratio should be slightly more negative than positive ions: 1.2 to 1. However, ions are highly unstable; in our industrialized, largely indoor lives, filled with electromagnetic charge from pollution and artificial sources, this ideal number is drastically diminished and the ratio disturbed, leaving all but the most robustly

outdoorsy among us inhaling too low a level of ions, with a predominance of positive ions. Living with low levels of ions is not particularly good for us – or for our ability as receivers or transmitters.
Research in California and Israel has shown that lower concentra-
tions of either positive or negative ions will produce fewer alpha
frequencies in the human brain and that sudden higher levels of
either charge can produce rapid, distinctive brain-wave changes.[50]

Persinger's research offers a vast amount of evidence that magnetic
frequency affects our ability to 'tune' in and transmit, and also
affects those portions of the brain that receive the information.
Subtle shifts in the earth's geomagnetic fields most noticeably affect
the heart and brain, the very systems of the body shown by the
DMILS research and Schlitz's Love Study to be the primary source
of transmission. After examining Persinger's work, I began to view
intention as a vast energetic relationship involving the sun, the
atmosphere, and earthly and circadian rhythms. To send intention
effectively, we would have to take account of these energies. Persinger
had usefully located not only the best 'channel' for intention, but also
the best time to turn it on.

CHAPTER EIGHT

The Right Place

IN 1997, WILLIAM TILLER had been helping a Californian company develop a product to eliminate electromagnetic pollution. The product contained a quartz crystal, which was why they had consulted him. Tiller, a physicist and professor emeritus of materials science and engineering at Stanford University, had carved out an influential niche for himself in the science of crystallization; he had written three textbooks on the subject and more than 250 scientific papers.[1]

The product consisted of a simple black box, about the size of a remote control. Inside its casing were three oscillators of 1–10 megahertz, barely a microwatt's worth of power when the device was turned on. The box also contained an electrically erasable, programmable, read-only memory (EEPROM) component, unconventionally connected in the circuit. It seemed to be able to screen incoming electromagnetic energy, possibly through the quartz oscillators also contained inside the box: quartz was thought to modulate quantum information by rotating the direction of waves.

As Tiller examined the equipment, an outrageous idea struck him. Fascinated by evidence that remote influence worked, Tiller had been carrying out a number of his own experiments and had formulated an entire theory about 'subtle energy' in living systems. Perhaps the little box he held in his hand might help him put intention to the supreme test. If thoughts were just another form of energy, what if he attempted to 'charge' this simple low-tech machine with a human intention and then use it to try to affect a chemical process? His experiment would rest on the unthinkable assumption that thoughts could be imprisoned in a bit of electronic memory and later 'released' to affect the physical world.[2] This fanciful idea would lead to a bizarre experimental result, offering convincing evidence that there is such a thing as the right place, as well as the right time, for carrying out intentions.

Tiller borrowed some lab space at the Terman Engineering building at Stanford from one of his tolerant colleagues in civil engineering, and some other space in the biology department, made some adjustments to the commercial device, and began designing his experiments. He wanted to go for broke, to see if this 'caged' intention could affect actual live test subjects. He realized he could not yet try his experiments on human beings, because they presented too many random, uncontrollable variables. But he could experiment on what scientists consider the next best thing to a human being: the fruit fly.

In the laboratory among the experimental animal population, the fruit fly is prom queen. Scientists have considered *Drosophila melanogaster* a model organism for more than a century, largely because its life cycle is so short. Within six days a fruit fly will completely remodel itself from larval grub to six-legged, winged insect and die just two weeks later. Tiller had in mind an experiment that would speed up their entire developmental process even further. His Stanford colleague Michael Kohane, an expert in fruit flies, had been studying the effects of supplements of nicotinamide adenine dinucleotide (NAD) on his fruit fly specimens. An important co-factor for enzymes, NAD helps in energy metabolism within cells by transporting hydrogen, which is essential in setting the fly's built-in timer for larval development. Energy availability also directly affects an organism's fitness.[3]

NAD marshals electrons into the pathway needed to maximize energy production and metabolism; low levels of NAD adversely affect the production of adenosine triphosphate (ATP). Every cell uses oxygen and glucose to convert ADP (adenine diphosphate) and phosphoric acid into ATP, a molecule that slow-drips energy to power most cellular processes. ADP and ATP are the equivalent of chemical energy storage tanks. Each molecule hoards a tiny supply of energy deep within its phosphorus–oxygen bond. Increasing the supply of NAD will increase the ratio of ATP to ADP, causing the cellular processes to work harder and faster, fast-forwarding larval development. As the fruit fly develops, the higher the ATP/ADP ratio, the more energy available to the cells, and the fitter the fly. The net effect of NAD is to increase a fruit fly's overall level of health, from cradle to grave.

Electromagnetic fields can have a profound effect on cellular energy metabolism, particularly the synthesis of ATP.[4] Human thoughts could be construed as a similar form of energy, Tiller reasoned. But could the energy of a thought interact with the transport chain of electrons to stoke up the metabolic fire?

To carry out the protocol he had in mind, Tiller needed a second lab. He set up one near the benefactor who was going to fund the studies in a small facility in Minnesota, just north of Excelsior. There he installed Michael Kohane and Walt Dibble, one of his former graduate students.

One morning in early January 1997, Tiller gathered his four participants, including himself, his wife Jean, and two friends, all highly experienced meditators, around a table. He unwrapped the first black box, placed it in the middle of the table and turned it on.

At the signal, Tiller told them all to enter a deep meditative state. After mentally 'cleansing' the environment and the equipment itself, he stood before them, a tall, lanky figure with bright, irreverent eyes and a wispy white beard, and read aloud the intention he had scripted earlier:

> Our intention is to synergistically influence (a) the availability of oxygen, protons, and ADP (b) the activity of the available concentration of NAD plus (c) the activity of the available enzymes, dehydrogenase and ATP-synthase, in the mitochondria so that the production of ATP in the fruit fly larvae is significantly increased (as much as possible without harming the life function of the larvae) and thus the larval development time significantly reduced relative to that with the control device.

Although the intention boiled down to significantly increasing the ratio of the ATP to ADP, Tiller had purposefully made the intention highly specific, so there would be no possible misunderstanding. He suspected that the more specific the thought, the more likely it was to have an effect, and so was careful, with each experiment, to pinpoint its aims. He had added 'without harming the life function of the larvae' because he suspected that if they tried to push things too far, they might well kill the tiny creatures.

The meditators held the intention for 15 minutes, before abruptly releasing it, at Tiller's signal, then they focused for a final 5 minutes on a closing intention, to mentally 'seal the intention' into the device.

Tiller had prepared an identical control box that had not been 'imprinted' with intention by wrapping it in aluminium foil and placing it in an electrically grounded Faraday cage, in order to screen out electromagnetic frequencies of all magnitudes.

He wrapped the imprinted black box, or the 'Intention-Imprinted Electronic Device', as he now called it, in aluminium foil and placed it in another Faraday cage until ready for shipping. On separate days he shipped each box via FedEx to the Minnesota laboratory, some

1500 miles away. He had been careful to blind the experiment so that neither Dibble nor Kohane would know which device contained the intention and which the control when the two devices arrived. The Excelsior scientists prepared several groups consisting of eight vials of fruit fly larvae and placed three of the groups of vials inside Faraday cages. They then placed both black boxes inside two of the cages with the vials and turned them on.

Over the next eight months, they carried out experiments on 10,000 larvae and 7000 adult flies, in each instance tracking the ATP/ADP ratio. After compiling their data and mapping it on a graph, Tiller and Kohane discovered not only that that the ratio of ATP to ADP had increased, but also that those larvae exposed to the imprinted devices developed 15 per cent faster than normal.[5] Furthermore, once the larvae had reached their adult stages they were healthier than normal, as were their descendants.[6] The intention not only had a positive effect on the flies themselves; it also appeared to affect the genealogical line.

By that time, Tiller had tried out other black boxes on a great number of other subjects, choosing his experimental targets with care. He needed tests like that of the fruit fly co-enzyme ratio that would show a genuine, measurable change. He decided on two new targets: the pH of water and the increase in the activity of a liver enzyme called alkaline phosphatase (ALP). He chose the pH test because water pH – the measure of acidity or alkalinity in a solution – remains fairly static and tiny changes of one-hundredth or even one-thousandth of a unit on the pH scale can be measured; a change of a full unit or more on the pH scale would represent an enormous shift that was unlikely to be the result of an incorrect measurement. ALP is another ideal test target because its activity proceeds at an unvarying rate.

In both instances, his meditators imprinted intentions into the black boxes to change the pH of water both up and down by a full pH unit and to increase by a 'significant factor' the activity of ALP. Tiller then sent off both imprinted and control boxes to Dibble, who made use of a similar study design as the fly experiment. Both experiments were extraordinarily successful.[7] In the water experiments, their intentions managed to change the pH up and down by one unit, and the ALP activity was significantly increased.[8]

Tiller was in the midst of some of his black-box experiments when he noticed something strange. After three months, the results of his studies began to improve; the more he repeated the experiment, the stronger and quicker the effects.

Tiller decided to try to isolate the aspect of the environment that had changed. He took readings of the air temperature, in and outside the Faraday cages, and discovered that the temperature appeared to be going up and down according to a regular rhythm or oscillation, dipping and climbing at regular intervals. He had first taken the temperature readings with an ordinary mercury thermometer. In case these results had something to do with the instrumentation, he switched to a computerized, low-resolution thermistor-based digital thermometer. Then he tried a high-resolution thermometer. All three recorded the same readings. When he plotted it, he saw that the temperature change was oscillating at a precise rhythm every 45 minutes or so, varying by some 4°C.[9] Tiller then measured the pH of water in the lab and measured its capacity to conduct electricity. He observed the same phenomenon as he had with the temperature: periodic oscillations of at least one-quarter of a unit on the pH scale, and regular dips and peaks in the water's ability to conduct electricity. Tiller was especially intrigued by the changes in pH. The acid/alkaline balance in any substance is highly sensitive to change; if the pH of a person's blood shifts up or down by just a half a pH unit, it means that they are dying or already dead.

A pattern was developing: as the temperature of the air rose, the pH fell, and vice versa, in near perfect harmonic rhythm. The water's electrical conductivity showed a similar harmonic cycle.[10]

Somehow his lab was beginning to manifest different material properties, almost as if it were a specially charged environment.

The effects also continually increased. No matter which experiment he carried out, the longer the imprinted devices were in the room, the larger the rhythmic fluctuations of the temperature and pH.[11] These fluctuations remained unaffected by the opening of doors or windows, the operation of air conditioners or heaters, and even the presence or movement of humans or objects around their immediate vicinity. When he compared graphs of air and water temperature readings, they again mapped in perfect harmony. Every corner of the room that was measured registered the same result. Each aspect of the physical space appeared to be in some sort of rhythmic, energetic harmony.

By this time, Tiller and his colleagues had set up four labs, each separated from the others by between 35 and 280 metres. Once enough experiments had been carried out, every other site also began to evidence these rhythmic fluctuations.

Tiller had never observed these types of 'organized' oscillations in his conventional science labs at Stanford. Indeed, they had never been observed anywhere else before. Just to be sure that this phenomenon was not being caused by the boxes themselves, he and his colleagues carried out three control experiments, in which devices that had not been imprinted with intention were placed in the spaces and turned on. In those cases, all the readings of air and water behaved normally.

Tiller still puzzled over the meaning of the effects, and whether they might be due to some physical disturbance. He wondered whether having two large fans in the room would affect the oscillations in the air and water. Ordinarily, forced air convection from a fan should cause oscillations in temperature to disappear. He placed a desk fan and a floor fan in strategic places near a line of temperature probes. Even when the fans were turned up high enough to scatter pieces of paper, the original temperature oscillations remained.

What exactly was going on? This could be a magnetic effect, Tiller thought. Perhaps he should check out the magnetic field of the water. He placed an ordinary bar magnet under a jar of water for three days, with the north pole of the magnet pointing upwards, and measured the water's pH. Then he turned the magnet over so that the south pole faced upwards under the jar for the same period. In normal circumstances, when ordinary water is exposed to this kind of weak magnet, which has a field strength of less than 500 gauss, the pH will be the same, no matter which side of the magnet is exposed to the water.

The world as we know it is magnetically symmetrical. Quantum physicists speak in terms of gauge theory and symmetry to explain the relationships between forces and particles, which include electric and magnetic charge. We are believed to exist in a state of electromagnetic $U\{1\}$-gauge symmetry – a rather complicated scenario in which magnetic force is proportional to the gradient of the square of the magnetic field. This boils down to a simple truism: no matter where in a given field you measure the electromagnetic property, you get the same reading. The electromagnetic laws of nature are the same wherever you look.

If you raise the electromagnetic pull in one area, you will find you have changed the electromagnetic pull by the same degree everywhere else. In *The Cosmic Code*,[12] Heinz Pagels likens the universe to an infinite piece of paper, painted grey. If you change the colour to a different shade of grey or 'change the gauge', you still don't change the gauge symmetry, because all the rest of the paper will be changed to the exact same shade of grey, so that it is even impossible to distinguish where exactly you are on the paper. A state of symmetrical magnetism is referred to as a magnetic 'dipole'.

But the pH of the water in Tiller's lab was significantly different with one polarity as compared with another, with huge differences of 1–1.5 pH units. Exposing the water to the south pole would send the pH soaring upward, while turning the magnet over to the north pole would cause the pH to decrease. At two of his experimental sites, the pH of the water, when exposed to the south-pole polarity, continued to change with the passage of time, peaking after about six days. When the water was exposed to the north pole of the magnet, however, the rhythmic changes in pH that he had been recording dwindled away.[13]

Orthodox science maintains that monopoles only exist in electricity (as a positive or negative charge), but not in magnetism, which creates only dipoles from spinning or orbiting electrical charge.[14] Governments around the world have spent billions of dollars looking for magnetic monopoles everywhere on earth, without success.[15] Somehow, Tiller had managed to access a magnetic monopole in his crude lab. This phenomenon appeared to be a system-wide effect. In any lab of his exposed to the intention-imprinted black boxes, instruments recorded magnetic monopole type of behaviour.

It dawned on Tiller that he was witnessing the most astonishing result of all: human intention captured in his little black boxes were somehow 'conditioning' the spaces where the experiments were carried out.

Tiller wondered whether this phenomenon would still be present if he altered anything about the space. When he removed one element, such as a computer, the oscillations disappeared for about 10 hours before returning. The arrival of any new materials in his lab also caused the effects to disappear for several weeks, although, once again, they eventually returned. It was as though the space had become an exquisitely tuned configuration, and no disturbance or change would destroy this higher state. Even when Tiller shielded the intention-imprinted devices in aluminium foil and Faraday cages, all the vibrations in water and air temperature continued. One of the sites, a converted barn, recorded oscillating air temperatures on and off for six months; in another site, an office lab, for a full year.[16]

After the imprinted boxes had been turned on for a while, the effect became relatively 'permanent'; the target, whether water pH,

ALP or fruit flies, would continue to be affected even if the device was not in the lab. Tiller decided to see what would happen when he removed all the elements of the experiment. He dismantled the Faraday cage and the water vessels and removed them from his lab, then recorded the air temperature of the place where the cages had been. Even though the experimental vessel was no longer there, his thermometers continued to record periodic oscillations in temperature of 2–3°C. Although this influence decayed very slowly over time, Tiller's laboratories appeared to have undergone some long-term thermodynamic transformation. The energy from intention appeared to 'charge' the environment and create a domino effect of order.[17]

The only other phenomena Tiller could think of that had similar effects on the environment were those of highly complex chemical reactions. But all he was working with was ordinary air and purified water. According to the laws of conventional thermodynamics, air and water are supposed to exist in a state extraordinarily close to equilibrium, which is to say that they remain more or less static. These types of results had never been recorded in any lab in the world.

He suspected that he had been witnessing a quantum effect. The constant replaying of ordered thoughts seemed to be changing the physical reality of the room, and making the quantum virtual particles of empty space more 'ordered'. And then, like a domino effect, the 'order' of the space appeared to assist the outcome of the experiment. Carrying out the intentions in one particular space appeared to enhance their effects over time.

Somehow, in these charged spaces he and his colleagues had managed to create a SU{2}-gauge space, where electric and magnetic monopoles coexisted – similar to the reality supposedly present in the supersymmetry states of exotic physics. In these conditioned spaces, the very law about the proportion of magnetic force had altered. *A basic property of physics had completely changed.* The only way to get such a polarity effect was to produce some element of SU{2}-gauge symmetry.[18]

This change in the gauge symmetry of the space meant that profound changes had occurred in the ambient Zero Point Field. In a U{1}-gauge symmetry, the random fluctuations of the Field have no effect on the physical universe. However, in SU{2}-gauge symmetry states, the Field has become more ordered and produces changes in the tiniest elements of matter – which add up to a profound alteration in the very fabric of physical reality.

Tiller felt as though he had somehow entered into a twilight zone of higher energy and that he was witness to a system with an extraordinary ability to self-organize. Indeed, the oscillations he had measured had all the hallmarks of a Bose–Einstein condensate – a higher state of coherence. Up until then, scientists had created a Bose–Einstein condensate only in highly controlled environments and at temperatures approaching absolute zero. But he had managed to create the same effects at room temperature, and from a thought process captured in a rudimentary piece of equipment.

Other scientists have witnessed a similar 'charging' of intention space. In one series of meticulous studies, for instance, researcher Graham Watkins and his wife Anita recruited human participants, many known for their psychic ability, and asked them to attempt to mentally influence anaesthetized mice to revive more quickly than usual. The experimental mice were drawn from a batch that had demonstrated similar waking times when placed under anaesthesia; the chosen group were divided in two, with half acting as controls.

In the first batch of studies, the experimental group woke up about 4 seconds earlier than the controls, a result considered only slightly significant. However, in subsequent studies the wake-up times of the experimental mice improved, and continued to do so with every study.

The Watkinses repeated their experiment seven times. They discovered that the healing had a 'linger effect'; if a mouse were simply placed on the spot on a table where another mouse had received a psychic's intentions, the second one would also revive more quickly than usual. The space appeared to have developed a healing 'charge', affecting anything that happened to occupy that space.[19]

Biologist Bernard Grad at McGill University in Montreal, Canada, observed a similar phenomenon during experiments with Hungarian healer Oscar Estabany: once the Hungarian healer touched something – even simple fabric – it appeared to hold a phantom charge. The material could be used successfully for healing in place of Estabany's healing hands.[20]

The idea of 'conditioned space' was also explored by former PEAR scientist Dr Roger Nelson at sacred sites. Nelson was intrigued by these sacred spaces and whether their special purpose, or even some inherent quality about the site, had 'charged' the space with an energetic resonance that might register on a REG machine. He had run a number of experiments suggesting that a 'field consciousness' in a highly charged atmosphere, such as an intense gathering, affected the machines and made them more 'ordered'. He carried around a portable REG, to register any changes in the randomness in the ambient field at various sites: Wounded Knee, the site of the massacre of an entire Sioux tribe; Devil's Tower in Wyoming; the Queen's Chamber in the Great Pyramid of Giza. Nelson registered highly significant evidence of increased order on REGs at some sites, as if the location itself contained a lingering vortex of coherent energy, from all the people who had prayed or died there.[21]

Dean Radin used REGs to investigate whether healing can condition the place where it is carried out. He placed three REGs near a culture of human brain cells, then asked a group of healers to send intentions for the culture to grow more quickly, and to engage in traditional space-conditioning meditations.

Any deviation from the random activity of the REGs would indicate the probable presence of greater coherence. Radin also prepared a control batch of cells, which were not to be sent intention.

After three days, there was no overall difference in the growth between the treated cells and the controls. Nevertheless, as the experiment progressed, the treated cells began to grow faster. On the third day, all three of the REGs began moving away from random activity and became more ordered. The intention of the healers also appeared to have effects on background ionizing radiation.[22]

Like Nelson's readings at sacred sites, Radin's experiment offers tantalizing hints about the nature of the 'linger' effect of intention. The REGs' registering of movement away from randomness to greater order implies that the Zero Point energy of empty space has shifted into a state of greater coherence. The 'charge' of intention may have a domino effect on its environment, causing greater quantum order in empty space, which would enhance the effectiveness of its aim.[23] Russian scientists have observed a similar phenomenon in water, which retains a memory of applied electromagnetic fields for hours, even days.[24] The effect is like that of a laser; when waves of the ambient Field become more ordered, an intention may ripple through it like one powerful, highly targeted bolt of light.

With magnetic monopoles, Tiller was out on a ledge shared by few of his colleagues, even in consciousness research. His studies needed to be replicated by other, independent laboratories. But if his body of work does stand up over time, it will demonstrate the extent to which the energy of human thought can alter its environment. The ordering process of intention appears to carry on, perpetuating, possibly even intensifying its charge.

The strange, almost unbelievable events occurring during Tiller's experiments made me wonder whether setting aside a particular room for carrying out intention might be an important consideration. Perhaps we each need our own 'temple' to which we return, if only in our mind's eye, every time we send a directed thought.

PART THREE

The Power of Your Thoughts

Baseball is 90 per cent mental.
The other half is physical.
Yogi Berra

CHAPTER NINE

Mental Blueprints

SEVEN WEEKS BEFORE MUHAMMAD ALI met World Heavyweight Champion George Foreman for their 'rumble in the jungle' at Kinshasa in 1974, he practised his punches as if he couldn't care less, taking a few desultory swipes at his sparring partner as if distractedly popping a speed bag. Mostly he would lie against the ropes and allow his opponent to pound away at him from every angle.

In the latter years of his fighting career, Ali spent much of his training time learning how to take punches. He studied how to shift his head by just a hair a microsecond before the connection was made, or where in his body he could mentally deflect the punch, so that it would no longer hurt. He was not training his body to win. He was training his mind not to lose, at the point when deep fatigue sets in around the twelfth round and most boxers cave in.[1] The most important work was being done, not in the ring, but in his armchair. He was fighting the fight in his head.

Ali was a master of intention. He developed a set of mental skills that eventually altered his performance in the ring. Before a fight, Ali used every self-motivational technique out there: affirmation; visualization; mental rehearsal; self-confirmation; and perhaps the most powerful epigram of personal worth ever uttered: 'I am the greatest'. Ali also made public statements of his intentions. His constant barrage of rhyming couplets and quatrains, seemingly so innocuous, were highly specific intentions in disguise:

Archie Moore
Is sure
To hug the floor
By the end of four

Now Clay swings with a right
What a beautiful swing
And the punch knocks the Bear
Clear out of the ring.

Before a fight, Ali repeated these little rhymes like a mantra – to the press, to his opponent, and even in the ring – until he himself accepted them as fact.

When they met in Kinshasa, Foreman was seven years younger than Ali and among the most savage fighters in the ring. Just two months earlier, he had left Ken Norton for dead with five blows to the head after only two rounds.

Nevertheless, in the weeks before the fight, when reporters pressed Ali about the two-to-one odds against him, Ali had rewritten the history of the Norton–Foreman fight, which he repeated, virtually verbatim, to every journalist who interviewed him.

'He's got a hard-push punch but he can't hit,' he would say, punching the air in front of the reporter's nose. 'Foreman just pushes people down. He just got slow punches, take a year to get there. You think that's going to bother me? This is going to be the greatest upset in the history of boxing.'[2]

Ali's intention came to pass in the jungle. He also made masterful use of intention to beat Joe Frazier in the Philippines later that year, in perhaps the most brutal and stunning display of boxing of all time.

This time, he created a voodoo doll. Ali turned his ferocious opponent into a tiny rubber gorilla, which he carried around with him in his top pocket, taking a swipe at it with his right from time to time for the television cameras: 'It's gonna be a thrilla and a chilla and a killa when I get the gorilla in Manila.' By the time Frazier entered the ring, he had been reduced in his own mind to something less than human.

Besides these verbalized intentions, Ali carried out mental intentions by rehearsing every moment of the fight in his head: the fatigue in his legs, the sweat pouring off his body, the pain to his kidneys and bruises on his face, the flash of the photographers, the exultant screams of the crowd, even the moment when the referee lifts his arm in victory against Frazier. He sent an intention to his body to win and his body responded by following orders.

To take intention out of the laboratory, I began to sift through the data from people or groups who were using intention successfully in real life. I wanted to study their techniques, the particular thought processes they underwent when sending intention, and would try to extrapolate from their experiences some tools that all the rest of us could use when sending intention. I was also curious about the extent of their mental reach – just how far people had been able to push their intentions.

The most instructive examples came from sports, not only from the greatest athlete of all time, but also from other elite sportsmen and women. Athletes of all varieties now routinely practise what is variously termed 'mental rehearsals', 'implicit practice'[3] or even 'covert rehearsal'. Focused intention is now deemed essential to alter and improve performance. Swimmers, skaters, weightlifters and football players employ intention to enhance their level of performance and consistency. It is even being used in leisure sports, such as golf and rock climbing.

Any modern coach of a competitive sport routinely offers training in some form of mental rehearsal, and often it is touted as the decisive element separating the elite sportsperson from the second-division player. National-level soccer players, for instance, are more likely to use imagery than those who remain at the provincial or local levels.[4] Virtually all Canadian Olympic athletes use mental imagery.

Psychologist Allan Paivio, professor emeritus of the University of Western Ontario, first proposed that the brain uses 'dual coding' to process verbal and non-verbal information simultaneously.[5] Mental practice has been shown to work just as well as physical practice for patterns and timing.[6] Paivio's model has been largely adapted to help athletes with motivation or in learning or improving a certain skill set.[7] The techniques involved in mental rehearsal have been exhaustively studied and written about in scientific literature and popular publications,[8] and their credibility was given a further boost in 1990, when the National Academy of Sciences examined all the scientific studies to date on these methods and declared them effective.[9]

Athletic mental rehearsal has been incorrectly considered synonymous with visualization. 'Visualization' implies that you observe yourself in the situation, as if watching a mental video featuring yourself or seeing yourself through another pair of eyes. Although this may be useful in other areas of life, visualizing oneself from an external perspective in a sports event can hamper athletic performance. Mental rehearsal also differs from positive thinking; happy thoughts on their own do not work in competitive sports.[10]

The most successful internal rehearsal involves imagining the sports event from the athlete's perspective *as though he or she is actually competing*. It amounts to a mental trial run – Ali imagining his right fist at the moment of impact on Frazier's left eye. The athlete envisages the future in minute detail *as it is unfolding*. Champion athletes forecast and rehearse every aspect of the situation, and the steps they should take to overcome any possible setbacks.

Tracy Caulkin used intention to land a third gold medal in the 1984 Olympics. Caulkin had already broken 5 world records and 63 American records, and at the age of 23 was considered the best American swimmer who had ever lived. All she needed to complete her trophy wall was a few Olympic golds.

At the time, electronic touchpads had just replaced stopwatches. Whereas the watch could only distinguish differences of hundredths of a second, the new electronic technology could distinguish the winning lead within a thousandth of a second – 400 times faster than the blink of an eye. In the Olympics relay swimmers are given two-hundredths of a second of grace to leave their block before their previous team mate hits the touchpad. This kind of fine timing is critical; even a single coat of paint on one side of the pool can make a swimmer's lane one-thousandth of a second longer to swim and give another swimmer the leading edge.

During the four-woman 400-metre relay race, Tracy took the lead by diving in one-hundredth of a second before her returning team mate hit the touchpad.

Although all her competitors had a similar level of fitness, Caulkin had one enormous advantage. She already knew every moment of her swim, from the dive and the cool rush of water past her head to the very moment when she would lunge out in front. Tracy had practised that hair's width lead, the precise moment when she would leave the block a hundredth of a second earlier than her opponents, every night inside her head. The conclusion of the Olympic relay had entirely depended upon the specificity of her intention.

The most successful athletes break down their performances into tiny component parts and work on improving specific aspects. For general mastery of their sport, they imagine a flawless performance.[11] They concentrate on the most difficult moments and work out good coping strategies – how to stay in control in the face of adversity, such as a pulled muscle or an umpire's adverse call. Different intention is employed, depending on whether they are first learning a skill or simply wishing to reinforce and improve their technique. Like Muhammad Ali, elite athletes learn how to block out images representing doubt. If an image of difficulty pops into their heads, they become extremely adept at changing the internal movie, quickly editing the scene to imagine success.[12]

Winning depends on how specifically you can mentally rehearse. Seasoned athletes use vivid, highly detailed internal images and run-throughs of the entire performance.[13] The most important aspect of the intention is to rehearse the victory, which appears to help secure it. Successful competitors rehearse their own feelings, particularly their elation and emotional response to winning: the reactions of their parents, the medals, the post-match celebration and the residual rewards like sponsorships.[14] They imagine that the crowd is cheering for their performance alone.

Experienced athletes engage all their senses in their mental rehearsal. They not only have a visual, internal image of the future event, they also hear it, feel it, smell it and taste it – the ambience, the competitors, the sweat of their bodies, the applause. Of all the sensations, the most vital for athletes appears to be mentally rehearsing the 'feel', or kinaesthetic sensations in their bodies.[15] The more experienced the athletes, the better they are at imagining the feel of their bodies when engaged in their sport.[16] Champion rowers are most successful when they can forecast the 'feel' of every part of the race, from the drag on the oar to the strain on their muscles.[17]

Some athletes find that it helps to study the actual setting where the sporting event is to take place first and then to imagine themselves there. Those who can combine the knowledge of the sports venue with mental rehearsal tend to be more successful than those who simply use mental rehearsal on its own.[18]

Rocky Bleier, former running back for the Pittsburgh Steelers, used intention to help the Steelers win the Super Bowl. His technique was to saturate his mind with the details of specific plays. He carried out mental rehearsals in the morning, before the team meal and last thing before drifting off to sleep every day of the two weeks before a game. He also found it reassuring to run through the entire catalogue of moves one final time just before play. While sitting on the bench, he again rehearsed some 30 runs and 30 passes. No matter what the field threw up to him that day, he was determined to be ready.[19]

Techniques differ among the various sports. Those mental rehearsals that work best for sports requiring aerobic ability and fast, coordinated movement tended to fail with strength training. Weight lifters are most successful after carrying out a mental intention that galvanizes them to lift an impossibly heavy object.[20]

Conventional wisdom has it that the best state for performance is a state of relaxation, but as I found with masters of intention, a relaxed state is not necessarily optimum. In a study of karate, using relaxation techniques before carrying out the intentions did not improve performance.[21] It was only useful if the participant was nervous and needed to be calmed down in order to perform better.[22] Relaxation and hypnosis used with intention have worked to improve aim – say, for basketball shots or accuracy in chipping in golf.[23] But as with Davidson's Buddhists, the most successful athletes manage to work themselves into peak intensity – a state of calm hyperawareness.

But how can simply thinking about a future performance actually affect the day of the event? Some clues come from intriguing brain research with electromyography (EMG). EMG offers a real-time snapshot of the brain's instructions to the body – when and where it tells it to move – by recording every electrical impulse sent from motor neurons to specific muscles to cause a contraction. Ordinarily,

EMG offers doctors a useful tool to diagnose neuromuscular disease and to test whether muscles respond appropriately to stimulation.

But EMG has also been employed to solve an interesting scientific conundrum: whether the brain differentiates between a thought and an action. Does the thought of an action create the same pattern in neurotransmission as the action itself? This very question was tested by wiring a group of skiers to EMG equipment while they were carrying out mental rehearsals. As the skiers mentally rehearsed the downhill runs, the electrical impulses heading to their muscles were just the same as those they used to make turns and jumps actually skiing the run.[24] The brain sent the same instructions to the body, whether the skiers were simply *thinking* of a particular movement or actually carrying it out. *Thought produced the same mental instructions as action.*

Research with EEGs has shown that the electrical activity produced by the brain is identical, whether we are *thinking* about doing something or actually doing it. In weightlifters, for instance, EEG patterns in the brain that would be activated to produce the actual motor skills are activated while the skill is simply being simulated mentally.[25] *Just the thought is enough to produce the neural instructions to carry out the physical act.*

Based on this research, scientists have posited some interesting theories of how mental rehearsal works. One school of thought proposes that mental rehearsal creates the neural patterns necessary for the real thing. As though the brain were simply another muscle, these rehearsals train the brain to facilitate the moves more easily during the actual performance.[26]

When an athlete performs, the nerves that signal to the muscles along a particular pathway are stimulated and the chemicals that have been produced remain there for a short period. Any future stimulation along the same pathways is made easier by the residual effects of the earlier connections. We get better at physical tasks because our signalling from intention to action has already been forged. It is not unlike a train track laid down through wild, inhospitable country. Future performances improve because your brain already knows the route and follows the track already laid down. Because the brain does not distinguish between doing something specific and just *thinking* about doing it, mental rehearsal lays down the tracks just as well as physical practice does. The nerves and muscles create a pathway just as sound as one produced through repeated practice.

Nevertheless, there are a few important differences between mental and physical practice. With physical practice, when you prac- tise too much, you become fatigued, which causes electrical interfer- ence and blockage along the tracks. With mental intention, no road blocks ever appear, no matter how much you practise in your head.

The other difference concerns the size of the effect; the neuromus- cular pattern laid down with mental practice may be slightly smaller than that of physical practice. Although both types of practice create the same muscle patterns, the imagined performances have smaller magnitude.[27]

To derive any benefit, mental rehearsal must replicate the real thing – at normal speed. Although it might seem logical that a rehearsal would work best in slow motion, with particular attention to specific moves, that is not borne out by research. When skiers monitored by EMGs imagined their performance in slow motion, they produced a different muscle response pattern from that produced when carrying out the skill at an ordinary pace. In fact, the brain–muscle activity of rehearsing at slow motion is identical to the brain–muscle pattern when the skiing itself is carried out in slow motion. This accords with what scientists understand about the neural patterns involved in slow motion, compared with those of normal speed. The same task carried out in slow motion produces completely different neuromuscular patterning than when it is done at normal speed.[28]

There is no such thing as cross-training in mental rehearsal; intention facilitates just the type of athletic event that is being mentally rehearsed and is not transferable to other sports, even those involving similar muscle groups. This was apparent in a fascinating study involving sprinters. The researchers had divided a group of runners into four groups and asked them to do one of four types of preparation: to imagine themselves in a 40-metre sprint; to engage in power training on a stationary bicycle; to combine imagery and power training; or, as the controls, to do no training in any form. After six weeks of training the athletes were asked to perform two tests – to cycle their hardest while their effort was recorded on a cycle ergometer, which tests for cycling power, and to run a 40-metre sprint. Both activities require much the same motor ability and leg muscles.

In the cycling test, those groups who had used power training alone showed improvement. However, when it came to the sprint, only the groups who had mentally practised sprinting had significantly improved. Specific imagery enhanced *only the specific task that had been imagined*. It did not simply build muscles generally. The motor neuron training was highly specific, and only affected the actual performance visualized in the mind.[29]

Beside improving performance, mental intention can produce actual physiological changes, and not only in athletes' bodies. Guang Yue, an exercise psychologist at Cleveland Clinic Foundation in Ohio, carried out research comparing participants who went to the gym with those who carried out a virtual workout in their heads. Those who regularly visited the gym were able to increase their muscle strength by 30 per cent. But even those who remained in their armchairs and ran through a mental rehearsal of the weight training in their minds increased muscle power by almost half as much.

Volunteers between 20 and 35 years old imagined flexing one of their biceps as hard as they could during daily training sessions carried out five times a week. After ensuring that the participants were not doing any actual exercise, including tensing their muscles, the researchers discovered an astonishing 13.5 per cent increase in muscle size and strength after just a few weeks, an advantage that remained for three months after the mental training stopped.[30]

In 1997, Dr David Smith at Chester College came up with similar results: participants who worked out could achieve 30 per cent increases in strength, while those who just imagined themselves doing the training achieved a 16 per cent increase.[31] Pure directed thought can give you the burn almost as well as any workout.

Thinking of changing an aspect of the body in other ways can also work – which might prove comforting to anyone who is not happy with his or her body shape. One study demonstrated that, under hypnosis, women increased the dimensions of their breasts simply by visualizing themselves on the beach with the sun's rays warming their chests.[32]

The kinds of vivid visualization techniques used by athletes are also highly effective in treating illness. Patients have boosted treatment of an array of acute and chronic conditions, from coronary artery disease[33] and high blood pressure to low-back pain and musculoskeletal diseases,[34] including fibromyalgia,[35] by using mental pictures or metaphoric representations of their bodies fighting the illness. Visualization has also improved postsurgical outcomes,[36] helped with pain management[37] and minimized the side-effects of chemotherapy.[38]

Indeed, the outcome of a patient's illness has been predicted by examining the types of visualizations used to combat them. Psychologist Jeanne Achterberg, who healed herself of a rare cancer of the eye through imagery, went on to study a group of cancer patients who were using visualization to fight their own disease. She predicted with 93 per cent accuracy which patients would completely recover and which would get worse or die, simply by examining their visualizations and rating them. Those who were successful had a greater ability to visualize vividly, with powerful imagery and symbols, and could hold a clear visual intention imagining themselves overpowering the cancer and the medical treatment being effective. The successful patients also practised their visualizations regularly.[39]

If the brain cannot distinguish between a thought and an action, would the body follow mental instructions of any sort? If I send my body a mental intention to calm down or speed up, will it necessarily listen to me? Literature about biofeedback and mind–body medicine indicates that it will. In 1961, Neal Miller, a behavioural neuroscientist at Yale University, first proposed that people can be taught to mentally influence their autonomic nervous system and control mechanisms such as blood pressure and bowel movements, much as a child learns to ride a bicycle. He conducted a series of remarkable conditioning-and-reward experiments on rats. Miller discovered that, if he stimulated the pleasure centre in the brain, his rats could be trained to decrease their heart rate at will, control the rate at which urine filled their kidneys, even create different dilations in the blood vessels of each ear.[40] If relatively simple animals like rats could achieve this remarkable level of internal control, Miller figured, couldn't human beings, with their greater intelligence, regulate more bodily processes?

After these early revelations, many scientists found that information about the autonomic nervous system could be fed back to a person as 'biofeedback' to pinpoint where a person should send intention to his body. In the 1960s, John Basmajian, a professor of medicine at McMaster University in Ontario and a specialist in rehabilitative science, began training people with spinal-cord injuries to use EMG feedback to regain control over single cells in their spinal cords.[41] At roughly the same time, psychologist Elmer Green at the Menninger Institute pioneered a method of biofeedback to treat migraine, after discovering that a migraine patient of his could make her headaches go away whenever she practised a structured form of relaxation. Green went on to use biofeedback to help patients cure their own migraines, and it is now an accepted form of therapy.[42] Biofeedback is particularly useful to treat Raynaud's disease, a vascular condition in which blood vessels are constricted when exposed to cold, causing extremities to grow cold, pale, and even blue.[43]

During a biofeedback treatment, a patient is hooked up to a computer. Transducers applied to different parts of his or her body send information to a visual display, which registers activities of the autonomic nervous system, such as brain waves, blood pressure and heart rate, or muscle contractions. The audio or visual information fed back to the patient depends on the condition; in the case of Raynaud's, as soon as the arteries to the hands constrict, the machines record a drop in skin temperature, a light bulb flashes or a beeper sounds. The feedback prompts the patient to send an intention to his body to adjust the process in question – in the case of Raynaud's, the patient sends an intention to warm up his hands.

Since those early days, biofeedback has become well established as a therapy for virtually every chronic condition, from attention deficit hyperactivity disorder (ADHD) to menopausal hot flushes. Stroke patients and victims of spinal-cord injuries now use biofeedback to rehabilitate or regain the use of paralysed muscles. It has proved invaluable in eliminating the pain felt in a phantom limb.[44] Astronauts have even used biofeedback to cure motion sickness while journeying to outer space.[45]

The more conventional view of biofeedback maintains that it has something to do with relaxation – learning to calm down the fight-or-flight responses of our autonomic nervous systems. However, the sheer breadth of control would argue that the mechanism has more to do with the power of intention. Virtually every bodily process measurable on a machine – even a single nerve cell controlling a muscle fibre – appears to be within an individual's control. Volunteers in studies have achieved total mental mastery over the temperature in their bodies,[46] or even the direction of blood flow to the brain.[47]

Like biofeedback, Autogenic Training, the technique developed by a German psychiatrist named Johannes Schultz to relax the body and slow the breathing and heart rate, also demonstrates that a wide variety of the body's functions are under our conscious control. Those who practise the technique are able to lower blood pressure, raise temperature in extremities, and slow heartbeat and breathing. Autogenic Training has also been used to treat many chronic conditions besides stress, such as asthma, gastritis and ulcers, high blood pressure and thyroid problems.[48] There is even evidence that Autogenic Training can work effectively in groups.[49]

For a cat, nirvana is the food bowl just around the corner. Dr Jaak Panksepp, professor emeritus of psychology at Bowling Green University, theorizes that this anticipatory joy has to do with the 'seeking' mode of the brain – one of the five primitive emotions that humans share with members of the animal kingdom.[50] The seeking system helps animals investigate and work out the meaning of their environment. The seeking circuits are fully engaged when an animal is involved in high anticipation, intense interest or insatiable curiosity. As Panksepp was astonished to discover, the most emotionally arresting part for any animal is the hunt, not the catch.[51]

When animals are curious, the hypothalamus lights up and the 'feel-good' neurotransmitter dopamine is produced. Scientists used to believe that the chemical itself caused the pleasure, until it was discovered that the chemical's true purpose is to arouse a certain neural pathway. What actually feels good is the activation of the seeking portion of the brain.

Forty years ago, Barry Sterman, professor emeritus of the departments of Neurobiology and Biobehavioral Psychiatry at UCLA, accidentally discovered that this anticipatory emotion sent cats into a meditative state; their brains slowed to an EEG rhythm of 8–13 hertz, corresponding to human alpha brain frequencies, moments before they got their reward.[52] Eventually, he was able to get the cats to re-create this state at will, not simply when they were awaiting food. It was tantamount to the animals being able to control their own brain waves.

But could a human being do the same? To test this, Sterman needed to test someone whose brain waves were so out of the ordinary that any change would be apparent immediately. He located a woman troubled by periodic epileptic seizures, which are caused by the brain firing theta brain waves at inappropriate moments. Sterman constructed a biofeedback EEG machine that would flash a red light in the presence of a theta wave and a green light during an alpha state. After a while, his patient was able to change her state at will and reduce the amount and intensity of her epileptic fits. Sterman spent the next 10 years of his life studying epileptics and training them to reduce their own fits.[53]

In the 1980s, two American psychologists, Eugene Peniston and Paul Kulkosky, made use of Sterman's findings to reform alcoholics. With their brain-wave biofeedback, alcoholic patients concentrated on damping down high beta brain waves, which tend to be predominant during moments of craving and dependency, and increasing the alpha and theta wave frequencies, which help one to relax and establish greater brain-wave coherence. Some 80 per cent of the alcoholics were able to control their cravings and stay off alcohol. The training also seemed to affect their blood chemistry, increasing their levels of beta-endorphin, another 'feel-good' brain chemical. Biofeedback, combined with work on their self-image, eventually eliminated much of their dysfunctional behaviour and transformed them into better people.[54]

Joe Kamiya, a psychologist at the University of Chicago, demonstrated the amazing specificity of brain-wave biofeedback through some remarkable brain research. He attached EEG electrodes to the rear sides of the scalps of several volunteers, over the portion of the brain where alpha brain waves are most prominent. At the sound of a tone, his participants had to guess whether their brains waves were predominantly alpha. After comparing their answers with the information recorded on the EEG machines, Kamiya let them know whether they were right or wrong. By the second day, one of his participants was able to guess correctly two-thirds of the time, and two days after that, virtually all the time. A second participant discovered a means of putting himself into a particular brain-wave state on cue.[55]

EEG biofeedback has now developed into a sophisticated means of controlling the range and type of frequencies emitted by the brain. It works particularly well with trauma patients suffering from depression,[56] helps students concentrate, and enhances creativity and focus. It may well be that intention can be used to control the brain, brain wave by brain wave.

Hypnosis is also a type of intention – an instruction to the brain during an altered state. Hypnotists continually demonstrate that the brain or body is susceptible to the power of directed thought.

One dramatic example of the power of mental suggestion concerned a small group of people with a mysterious congenital illness called ichthyosiform erythroderma, known disparagingly as fish-skin disease because unsightly fish-like scales cover most of the body. In one study, five patients were hypnotized and told to focus on a part of their body and visualize the skin becoming normal. Within just a few weeks, 80 per cent of each patient's body had completely healed. The skin remained smooth and clear.[57]

Through hypnotic intention, spinal-surgery patients about to undergo their operations have reduced blood loss by nearly half, simply by directing their blood supply away from the site of the surgery.[58] Pregnant women have been able to turn their babies from breech positions, burn victims have sped up their healing; and people suffering haemorrhages in the gastrointestinal tract have willed their bleeding to stop.[59] Clearly, during an altered state, roughly corresponding to the hyperalert state of intense meditation, conscious thought can convince the body to endure pain, cure many serious diseases and change virtually any condition.

Surgeon Dr Angel Escudero of Valencia, Spain, has carried out more than 900 cases of complex surgery without anaesthesia. BBC cameras were invited into his operating room and captured on film a woman who was having such an operation without anaesthetic. All she had to do was keep her mouth full of saliva and keep repeating to herself, 'My leg is anaesthetized.' An affirmation like hers is another form of intention. A dry mouth is one of the mind's first warning signals of danger. When the mouth is kept lubricated, the brain relaxes, assumes all is well, and turns off its pain receptors, assured that anaesthetics had been given.[60]

A fascinating study by David Spiegel, a professor of psychiatry and behavioural sciences at Stanford University, offers a glimpse of what happens to the brain when an intention is given under hypnosis. His participants were shown a coloured grid painting, similar to a Mondrian, and were asked to imagine the colour draining from the picture, leaving only black and white. Through the use of positron emission tomography (PET) scans, which record physical activity in the brain, Spiegel showed that blood flow and activity were noticeably diminishing in the part of the brain dealing with the perception of colour, while the areas that process black, white and grey images were being stimulated.

When the experiment was reversed, and the participants in the study were asked to imagine grey images turning into colour, the opposite changes in brain-perception patterns resulted.[61]

This illustrated another instance in which the brain was the maidservant of thoughts. The brain's visual cortex, the area responsible for processing images, could not distinguish between a real image and an imagined one. *The mental instructions were more important than the actual visual image.*

The placebo effect has shown that beliefs are powerful, even when the belief is false. The placebo is a form of intention – an instance of intention trickery. When a doctor gives a patient a placebo, or sugar pill, he or she is counting on the patient's belief that the drug will work. It is well documented that belief in a placebo will create the same physiological effects as that of an active agent – so much so that it causes the pharmaceutical industry enormous difficulty when designing drug trials. So many patients receive the same relief and even the same side effects with a placebo as with the drug itself that a placebo is not a true control. Our bodies do not distinguish between a chemical process and the *thought* of a chemical process. Indeed, a recent analysis of 46,000 heart patients, half of whom were taking a placebo, made the astonishing discovery that patients taking a placebo fared as well as those on the heart drug. The only factor determining survival seemed to be belief that the therapy will work and a willingness to follow it religiously. Those who stuck to doctor's orders to take their drug three times a day fared equally well whether they were taking a drug or just a sugar pill. Patients who tended not to survive were those who had been lax with their regime, regardless of whether they had been given a placebo or an actual drug.[62]

The power of the placebo was best illustrated by a group of patients treated for Parkinson's disease, a motor system disorder in which the body's system for releasing the brain chemical dopamine is faulty. The standard treatment for Parkinson's is a synthetic form of dopamine. In a study at the University of British Columbia, a team of doctors demonstrated with PET scanning that, when patients given placebos were told they had received dopamine, their brains substantially increased the release of their own stores of the chemical.[63] In another dramatic instance, at Methodist Hospital in Houston, Dr Bruce Moseley, a specialist in orthopaedics, recruited 150 patients with severe osteoarthritis of the knee and divided them into three groups. Two-thirds were either given arthroscopic lavage (which washes out degenerative tissue and debris with the aid of a little viewing tube) or another form of debridement (which sucks it out with a tiny vacuum cleaner). The third group were given a sham operation: the patients were surgically prepared, placed under anaesthesia and wheeled into the operating room. Incisions were made in their knees, but no procedure carried out.

Over the next two years, during which time none of the patients knew who had received the real operations and who had received the placebo treatment, all three groups reported moderate improvements in pain and function. In fact, the placebo group reported better results than some who had received the actual operation.[64] The mental expectation of healing was enough to marshal the body's healing mechanisms. The intention, brought about by the expectation of a successful operation, produced the physical change.

Extreme instances of intention and expectation can also manifest physically. The phenomenon of stigmata, in which religious fervour produces blood, bruising or wounds on people's hands, feet and sides that mirror the wounds of Christ during his crucifixion, are a form of intention. The Association for the Scientific Study of Anomalous Phenomena has recorded at least 350 such instances of stigmata resulting from identification with Christ. Saybrook University psychologist Stanley Krippner and his colleagues witnessed this first hand with Brazilian sensitive Amyr Amiden. As soon as their talk turned to Jesus Christ, red spots and drops of blood appeared on the backs of each of Amiden's hands and on his palms and forehead.[65] A similar situation occurred during the three weeks before Easter Sunday with a young African-American Baptist girl, who had been profoundly moved by a television movie about the crucifixion and was preoccupied with Christ's suffering. She manifested bleeding on the palm of her left hand two to six times a day.[66] Krippner knew of three Anglicans who regularly evidenced stigmata.[67]

Cases of spontaneous cures are an instance of an extreme intention that reverses almost certain death. A person with what is considered a terminal illness defies the textbook description of his disease progression and the prognoses of his doctors and beats it virtually overnight, without the aid of the tools of modern medicine.

The Institute of Noetic Sciences has gathered together all scientifically recorded cases of so-called miracle cures.[68] Although the received wisdom is that these cases are rare, a scan of the medical literature is instructive. One in eight skin cancers spontaneously heal, as do nearly one in five of genitourinary cancers. Virtually all types of illnesses, including diabetes, Addison's disease and atherosclerosis, where vital organs or body parts are supposedly irretrievably damaged, have spontaneously healed.[69] A small body of research concerns terminal cancer patients, who with little or no medical intervention, end up beating the odds.

Although these cases are labelled instances of 'spontaneous remission', as though the illness has suddenly decided to go into hiding but might suddenly spring out at any moment, in many instances they represent another example of the body's ability to self-correct through the power of intention. Case after case of spontaneous remission describes people up against a major road block in their lives: unremitting stress, unresolved trauma, prolonged hostility, marked isolation, profound dissatisfaction or quiet despair.[70] They often describe people who have lost their role as the central protagonist in their own life drama.[71]

Many cases of spontaneous remission seem to occur after someone makes a massive psychological shift, and recreates a life that is engaging and purposeful. In these instances, the patient gets rid of the source of the psychological heartache[72] and takes full responsibility for his illness and treatment.[73] Some people, this would suggest, get ill because they lose all hope of life ever being good – because they are thinking the wrong thoughts. These cases of spontaneous remission suggested to me that casual thoughts that run through our minds every day together become our life's intention.

We can use intention to gain control over virtually any bodily process and perhaps even life-threatening illnesses. But can our thoughts about others be as potent as our thoughts about ourselves?

Psychologist William Braud is one of the few scientists who has examined this question. He gathered a group of volunteers and asked them to carry out biofeedback on themselves. After pairing off the group, he attached one member of each pair to the biofeedback equipment, but asked the other partner to respond to the readings and carry out the sending of mental instructions. According to Braud's evidence, the results were equivalent to those that occurred when the patients on the equipment used biofeedback on themselves. Somebody else's good intentions for you may be as powerful as your own.[74]

Braud's other studies also suggested that we can most influence others to become more 'ordered' when we ourselves are ordered. For instance, in his studies, calm people were the most successful at sending mental influence to calm down highly nervous people, and focused people the best at helping distracted people focus.[75] Braud's work also suggests that the greatest effects occur when the person most needs help.[76]

Scientific evidence also reveals that we can affect virtually any other living thing as well. The enormous body of research on healing gathered by Dr Daniel Benor shows that thoughts can have powerful effects on a variety of plants, seeds, single-celled organisms such as bacteria and yeast, and insects and other small animals.[77] Most recently, a series of double-blind experiments carried out over two years by Dr Serena Roney-Dougal in Somerset showed that lettuce seeds that were sent intention yielded 10 per cent more crops with significantly less fungal disease than those grown conventionally.[78]

The evidence convinced me that we can improve our health, enhance our performance in every area of our lives, and possibly even affect the future by consciously using intention. The intention should be a highly specific aim or goal, which you should visualize in your mind's eye as having already occurred while you are in a state of concentrated focus and hyperawareness. When you imagine this future event, hold a mental picture of it as if it were occurring to you at that moment. Engage all five senses to visualize it in detail. The centrepiece of this mental picture should be the moment you achieve the goal.

A doctor might improve the survival rate of his patients by never giving a negative diagnosis.[79] A surgeon could improve his patients' recovery by mentally rehearsing the surgery before heading into the operating theatre. Indeed, we might no longer need drugs, but simply good intentions. Since intention has been shown to affect the chemistry in our bodies, we should be able to speed up, slow down or improve any physiological processes. We might develop many more breakthrough medicines by mentally targeting their effectiveness and minimizing their side effects.

We could raise the quality of our daily endeavours just by carrying out a detailed mental rehearsal. At home, we might be able to send intentions to our children to perform better at school or be more loving to their friends. Human intention might be powerful enough to affect every element of our lives.

All of these possibilities suggest that we have an awesome level of responsibility when generating our thoughts. Each of us is a potential

Frankenstein, with an extraordinary power to affect the living world around us. How many of us, after all, are sending out mostly positive thoughts?

CHAPTER TEN

The Voodoo Effect

DICK BLASBAND WAS DRAWN TO THE IDEA that there might be a way to amplify and direct life energy, like holding up a magnifying glass to focus the rays of the sun. Blasband, a psychologist, was intrigued by the theories of Wilhelm Reich, the Austrian psychiatrist and one-time protégé of Sigmund Freud, who thought it possible to trap 'orgone' – the name he gave to what he believed to be omnipresent cosmic energy – in an orgone energy 'accumulator'. An accumulator, a box-like enclosure of any size, could be made of alternating layers of any metal and non-metallic materials, such as cotton cloth or felt. Reich believed that atmospheric energy would be first attracted, then instantly repelled by the metal and eventually absorbed by the non-metallic substance. Because the enclosure was layered, orgone energy would continuously flow between the atmosphere and the box, like a current of air, and so constantly 'accumulate'. Reich had early encouraging results with animals and plants placed in the boxes, which lay the basis for his later claims that accumulated energy had an immense capacity to heal.

It occurred to Blasband that Reich's ideas about energy fields were not dissimilar to those of his colleague Fritz-Albert Popp and his work on biophotons. Perhaps the best means of testing an accumulator was to measure its effect on the emission of these tiny specks of light from a living thing.

In August 1993, Blasband travelled to Popp's International Institute, then in Kaiserslautern, Germany. He and Popp created a variety of orgone accumulators, then chose a number of plants in Popp's laboratory – cress seeds, cress seedlings and *Acetabularia crenulata*, a primitive form of marine algae – to be the experimental population. Popp's photomultipliers would count the light emissions of all the plants inside and outside the orgone boxes and record any differences.

Blasband carried out four experiments – placing the algae in the accumulator first for one hour, and then continuously for two weeks – with no result. Popp's equipment recorded not the slightest alteration in the light emissions. Blasband wondered if this was because the plants were already so healthy that the boxes could not improve their state of health. Perhaps he would see a larger change in a subject that needed help or improvement. He and Popp decided to try making the *Acetabularia* 'ill' by depriving it of most of its vitamin supply for 24 hours before treatment. It appeared to make no difference. The plants' biophotons didn't change. No amount of exposure to an accumulator of any variety seemed to make one bit of difference to the health or well-being of any of the plants.

Blasband and Popp then decided to test whether a mental intention could boost the action of the accumulators. In his new series of experiments, Blasband sent an intention for the energy within the accumulator to be beneficial to certain seedlings and harmful to others. These results were disappointing, too. There was only one significant difference in the number or quality of biophoton emissions before and after treatment of any of the plants: the only effective intention appeared to be the one he had sent to stunt their growth.[1] In both experiments, negative intention was more powerful than positive intention. Thoughts to harm had the greatest effect.

Blasband's little study highlights perhaps the most disturbing consideration of all about intention: that bad thoughts, as well as good ones, can have an effect on the world, and indeed may be the more powerful of the two. After all, in many native cultures, prayer and intention have a shadow component in hexes, voodoo and curses, which are reported to be highly effective forms of negative intention.

Many healers routinely use a negative means to a positive end. As Dr Larry Dossey, author of *Be Careful What You Pray For …*[2] has noted, negative intention is the very foundation of most healing. Healing from an infectious agent or a rogue cell line such as cancer requires intent to harm.[3] It works from a desire to *kill* something: to inhibit bacterial enzymes, alter cell membrane permeability, or interfere with the nutrition given to the cell or the synthesis of DNA.[4] In order for the patient to get better, the offending agent has to die.

Many pioneers of mind–body medicine in the treatment of cancer, such as Dr Bernie Siegel, Dr Carl Simonton, and Australian psychiatrist Ainslie Meares, encouraged their patients to use vivid forms of mental imagery – a metaphoric representation of their illness – to enhance their healing.[5] The majority of the cancer patients who first made use of visualization techniques imagined a battlefield, on which good (the patient) is pitted against evil (the cancer), with the cancer patient possessing the larger weapon. Some patients imagined their white blood cells as an army killing the cancer cells or a 'tap' containing the blood that feeds the cancer cells, which they can turn off. Some patients visualized themselves as participating in a violent video game. When Simonton first introduced this technique to his patients in the 1970s, Pac-Man was the most popular video game of the time. He encouraged his patients to imagine a little Pac-Man inside their bodies, gobbling up cancer cells in its path. But whatever the particulars of the imagery, the intention itself needed to be murderous; the patient had to want to annihilate the enemy.

Research on negative mental influence presents a number of obstacles to scientists. One basic problem, as Cleve Backster found in his research, is finding a living thing that no one objects to having killed. Many choose to study the most basic life forms, such as paramecia or fungi, or to experiment with seeds or small plants.[6]

Another problem is avoiding an unintended 'spray' effect: what if a healer's aim is slightly off one day and the negative intent gets sent to the host instead? The Canadian healer Olga Worrell refused to carry out negative intention on infectious diseases for exactly that reason. She worried that her negative intent might move beyond the bacteria and accidentally target the person she was trying to heal.[7]

One of the earliest experiments using negative intention was conducted by Jean Barry, president of the Institut Métapsychique International, who studied bacteria and fungi. As insignificant as these lowly organisms appear, Barry, a general practitioner, understood their pivotal role in maintaining health and causing illness. If it could be shown that intention has the power to eliminate these small organisms, humans might be able to exert greater control over their own health.

Barry decided to test the effect of negative intention on a fungus called *Rhizoctonia solani*. *Rhizoctonia*, a thready filament and a distant relative of the common mushroom, is an enemy to 500 types of crops. Farmers call it pod rot or root rot, as it commonly attacks the pods and roots, stunting growth and eventually consuming the plant. No one would object to a means of controlling this garden menace.

Barry set up a batch of experimental Petri dishes and matched them with a set of controls of the identical type of fungus growing in the same conditions. He enlisted ten volunteers and assigned five experimental Petri dishes and five controls to each person. At the appointed time, each volunteer was asked to send intentions to slow the growth of the fungi in the experimental Petri dishes. After the experiment, the lab assistant measured the growth of each sample of *Rhizoctonia* by outlining its boundary on tracing paper. Of the 195 dishes involved in the negative intention, 151, or 77 per cent, were smaller than the average size of the controls.[8]

Barry's study was successfully replicated by researchers from the University of Tennessee, although their study also tested the effect of remote influence; this time, the volunteers sending the intention were 15 miles away from the fungus samples.[9]

Similar research was conducted by Carroll Nash, the director of the parapsychology department at St Joseph's University in Philadelphia, but on *Escherichia coli*, microbes with a direct impact on human beings. Millions of these bacteria, which help to digest food and keep hostile bacteria at bay, peacefully reside in the gut. *E. coli* also metabolizes lactose, the enzyme present in milk. Yet, as with many microbes, *E. coli* can suddenly turn unfriendly by migrating out of the digestive tract or mutating into a virulent form that causes illness. Many toxic strains are also present in food. *E. coli* represented an interesting choice for Nash. If humans could control its growth, they might avoid serious *E. coli* infections and improve their general digestive health.

Nash decided to test whether mental influence could affect the mutation rates of *E. coli* bacteria. Usually, an *E. coli* population starts life unable to ferment lactose (and so is 'lactose-negative'), but after it mutates, over numerous generations, the new population can do so (at which point it become 'lactose-positive'). This process ordinarily occurs at a predictable rate. Nash wanted to see whether his volunteers could slow it down or speed it up. To work out the growth rates of these tiny organisms, Nash employed an electrophotometer, which counts the microbes by measuring the slightest differences in the density of the media in which they are suspended.

Each of his 60 student participants received nine test-tubes containing both lactose-negative and lactose-positive strains of *E. coli* culture. The students were asked to mentally encourage the transformation of the unmutated bacteria in the first three test-tubes from lactose-negative to lactose-positive. With the next three test-tubes, they were to attempt to inhibit the process of mutation. The final three, the controls, would not be exposed to influence of any kind. When he tallied the results, Nash discovered more mutation than normal in the test-tubes that had received the positive intentions to mutate, and fewer than normal in those for which the intentions were to inhibit the process, although the greatest effect occurred with negative intention.

Nash's study had an interesting coda: he had not stipulated any particular location where the mental influence had to originate; the volunteers were allowed to send their thoughts from the place of their choosing, whether the lab or elsewhere. When Nash examined the differences in the results according to the place from where the intentions had been sent, an interesting pattern emerged. Those students assigned the task of sending positive intentions had the best success if they sent their thoughts while in the lab; those with negative intentions had the best result if they waited until they had left. The Tennessee researchers who replicated Barry's study also discovered that negative intention was most effective when it was sent from a remote site. Positive intent appeared to work best in the presence of its object, whereas a negative intent worked best when the object of ill will was not anywhere in the line of sight.[10]

These early studies revealed several important aspects of intention. Thoughts take aim with great accuracy; although their effects on living things can drastically differ depending on the nature of the intention – whether it is positive or negative. Where we position ourselves when sending out a thought might also have a bearing on our success. Being near the target while sending a positive intention or away from the target when sending a negative intention might magnify its effect.

* * *

The next best experimental subject to a live human being is its cells. If you can prove an effect on an essential component of a living thing, it is likely that the same effect will occur with the entire organism. Dr John Kmetz, a colleague of William Braud's in San Antonio, Texas, decided to test the effect of negative intention on cancer. Although he could not test his theory on a live human being, he settled for a sample of cervical cancer cells, and enlisted Matthew Manning, a gifted British healer.

Manning sent negative intentions either by touching the beaker of cells or from a distance, inside an electromagnetically shielded room. Kmetz then used special equipment to count how many cancer cells were in the culture medium. Ordinarily, a cancer cell, which has a positive charge, will grasp the side of a plastic beaker, attracted to its negative electrostatic charge. An injury to the cell will cause it to drop off the side and into the culture medium. Kmetz's equipment demonstrated that Manning had fatally injured the culture.[11] Manning's extraordinary healing ability had been turned on its head; in this study he had become a killing machine.

Practitioners of *Qigong* openly acknowledge that intention has the power to enhance or destroy – indeed, the Chinese term for sending positive *Qi*, or life energy, through intention translates into English as 'peaceful mind', while sending negative *Qi* is referred to as 'destroying mind'.[12] A host of studies of *Qigong* carried out in China have been collated on the *Qigong* Database®, many of which claim to offer evidence that 'destroying mind' can kill human cancer cells or tumours in mice, decrease the growth rate of *E. coli* and inhibit activity of amylase, a digestive enzyme used to help digest carbohydrates.[13] Nonetheless, some Western scientists maintain quiet reservations about the database; few of these studies have been replicated in the West.

One study on plants conducted at the First World Conference of Academic Exchange of Medical *Qigong*, in Beijing in 1988, examined whether sending *Qi* could affect the growth of a confederate spiderwort plant by concentrating on its process of replication. A *Qi* master was asked to damage one of the plant's self-destruct mechanisms, which would cause the plant to live longer than normal.[14] The master had to target his negative intention precisely, so that it would injure only one aspect of the plant while the rest would thrive. To record any subtle effects on the health of the plant samples during the experiment, any increases or decreases in certain cells after replication, the researchers used a micronuclear method developed at Western Illinois State University. During the study, the *Qigong* master displayed a remarkable ability to send precise instructions to specific parts of the plant, some of which were damaging, some beneficial.[15]

A similar study was carried out by researchers at the National Yang Ming Medical College and National Research Institute of Chinese Medicine in Taipei, Taiwan. In this instance, the *Qigong* master alternately sent positive and negative intention to boar sperm cells and human fibroblast cells, which make up the connective tissue of the body. After 2 minutes of negative intention, the growth rates and protein synthesis of the cells decreased dramatically by 22–53 per cent. When the *Qigong* master reversed his intention and sent 10 minutes of positive intention, all the activity of the cell increased by 5–28 per cent.[16] In another well-controlled study by the Mt Sinai School of Medicine, two *Qigong* masters were able to inhibit the process involved in the contraction of muscles by as much as 23 per cent.[17]

These studies raise the obvious question: which is more powerful, a positive or a negative thought? In some studies, the will to harm appears to be the stronger of the two intentions, but that makes sense in a study like Blasband's, where it is probably far easier to damage a healthy system than to make a healthy system even healthier, or indeed to fix something that is broken, or to order a disordered system.[18] Nevertheless, effective intention of any variety is likely to require order and deliberately focused thought. How many negative intentions are sent by someone as ordered as a *Qigong* master?

Although negative intention appears capable of disrupting the most fundamental biological processes when precisely targeted,[19] one study suggests that healing does not necessarily require negative intention. Leonard Laskow, an American gynaecologist and healer, was recruited by American biologist Glen Rein to test the most effective healing strategy for inhibiting the growth of cancer cells. In his own practice, Laskow believed in establishing an emotional connectedness with his subject – even with cancer cells – before sending out healing. Rein prepared five different Petri dishes containing identical numbers of cancer cells and then asked Laskow to send out a different intention while holding each one. Laskow's first intention was that the natural order be reinstated and the cells' growth rate return to normal.

With the next Petri dish he was to adopt a Taoist visualization that entails imagining that only three of the cancer cells remained in the Petri dish. For the third dish he was not to have an intention, but simply to ask God to have His will flow through Laskow's hands. He offered unconditional love to the cancer cells of the fourth dish, which involved meditating on a state of love and compassion, much as Davidson's Buddhists had done. For the final dish of cancer cells, Laskow carried out his only truly destructive intention, by visualizing the cells dematerializing, either going into the light or the 'void'. Rein gave Laskow a choice of imagery largely because he was uncertain which visualization would be most effective in obliterating something. Was it more effective to release an entity by offering it an endpoint (the light), or simply to give it a full range of potential (the void)? As a yardstick of Laskow's effectiveness, Rein would measure the amount of radioactive thymidine absorbed by the cancer cells – an indicator of the growth rate of malignant cells.

Laskow's various intentions had quite different effects. The most powerful were undirected intentions asking the cells to return to the natural order, which inhibited the cancer cells' growth by 39 per cent. Acquiescing to God's will with no specific request was about half as effective, inhibiting the cells by 21 per cent, as was the Taoist visualization. An unconditional acceptance of the way things were had no effect either way, nor did imagining the cells dematerializing. In these two instances, the problem may have been that the thought was simply not focused enough.

In a follow-up study, Rein asked Laskow to limit himself to two possibilities, the Taoist visualization and a request for the cells to return to the natural order. This time, he achieved an identical result with both intentions; the cancer cell growth was inhibited by 20 per cent. The strongest effect of all occurred when he combined the two approaches, mixing an intention to return to the natural order while imagining only three cells left; his rate of cell inhibition doubled, to 40 per cent. Clearly the combination of asking the universe to restore order while imagining a specific outcome exerted a powerful effect. Rein asked Laskow to repeat this combined approach, but to target the medium in which the cancer cells grew, rather than the cells themselves. Laskow achieved the same result as when he had focused directly on the cells themselves.

Finally, Rein instructed Laskow to hold each of his five states of mind in turn while grasping one of five vials of water, which would later be used to make up the tissue-culture medium of the cancer cells. The water treated with the 'natural-order' intention again had the greatest effect, inhibiting the growth of the cancer cells by 28 per cent. In this case, water apparently 'stored' and transferred the intentions to the culture medium and on to the cancer cells.

Laskow's approach was instructive. The most effective healing intention had been framed as a request, combined with a highly specific visualization of the outcome, but not necessarily a destructive one.[20] With healing, the most effective approach may not be to destroy the source of the illness, but, as with other forms of intention, to move aside, let go of the outcome, and allow a greater intelligence to restore order.

* * *

Most research about negative intention concerns a conscious desire to harm something. I wondered about those moments when negative intention is unconscious. Suppose you don't like someone and harbour an unconscious ill will towards him? Do you unwittingly send out a negative intention? Or, what about those moments when you explode in anger? Is it possible that your momentary anger causes unintentional harm?

An overenthusiastic cleaner of mine once accidentally stripped off all the chrome on every fixture in our bathrooms. When I discovered the damage, a few hours after she had left our house, I was so over-whelmed with anger that I had to lie down. I had only just finished a five-month long renovation project on our newly purchased family home and had lovingly overseen the entire project, which had cost a good deal of hard-earned savings. I later learned, to my horror, that at about the time I had given voice to my fury, she had fallen off the bus and broken her leg. At another time, I was irrationally overwhelmed with anger at our bank manager, after discovering that our bank, now run by computers, had not recorded a deposit and had bounced several of our cheques. Later, I was horrified to learn that at roughly the moment I had vented my spleen, she had tripped on a pavement and broken most of her front teeth.

I had always felt guilty and curious about both these incidents. Was their misfortune my doing? Was it possible to curse people through your thoughts? I considered the effect of the everyday negative thoughts that swim through everyone's mind every day. A negative thought about yourself ('I'm untalented and lazy') or your children ('He's such a slob'; 'She's lousy at maths') might ultimately manifest as a physical energy and become a self-fulfilling prophecy. Indeed, moments when you feel an aversion to someone or something that you cannot rationally explain may simply be an instance when you are picking up a negative intention towards you. Even times when you are depressed could have a physical effect on the people and other living things around you.

Bernard Grad, the Canadian biologist, addressed many of these issues in a study that tested the power of a negative frame of mind on the growth of plants. He planted four groups of 18 pots, each containing 20 barley seeds. Each pot was to be watered with 1 per cent saline solution, slightly stronger than the kind used by hospitals when giving intravenous infusions to patients, which can stunt a plant's growth. Three batches of the plants were to receive watering with the salt water, but only after the water had been held by one of three people for half an hour. The control batch would be watered with the solution that had not been exposed to anyone.

The first vial was held by a healer with green fingers and a passion for plants. The other two vials were held by two depressed patients – a man diagnosed as a psychotic depressive and a woman who was neurotically depressed – chosen from the Canadian hospital where Grad worked. The man was so depressed that he didn't even ask what was in the bottle, but simply assumed that Grad, who wore a white coat, was just another in the procession of doctors preparing him for periodic electric shock therapy. While holding the bottle, he repeatedly protested that he didn't need an ECT treatment. The woman, on the other hand, visibly lifted when Grad told her that the bottle was part of an experiment. Half an hour later, when he came to retrieve the bottle from her, he discovered that she had been cradling it as if it were a baby.

This unforeseen turn of events worried Grad, as the woman had been chosen precisely because he believed she would be in a negative state of mind. She had suddenly appeared to regain her *joie de vivre*, simply at the thought of her involvement in the experiment. After carefully creating a multi-blind system so that he could not know or be influenced by who had done what, Grad poured the water over the seeds.

Several weeks later, he was pleased to see that the result more or less followed his prediction. The plants watered by the man with the psychotic depression grew the slowest, followed by the control plants, whose bottle had not been held by anyone. The fastest grow-ing plants had been watered by the green-fingered healer, followed by those of the depressed woman, which was a surprise. It seemed that her plants had grown faster because of her own enthusiasm about the experiment.[21]

Carroll Nash tried a similar experiment, asking a group of psychotics to hold individual sealed glass bottles of a solution of dextrose and sodium chloride for half an hour. Nash then removed 6 millilitres of the solutions from each bottle and poured them into fermentation tubes. Similar solutions that had not been 'charged' by the psychotics were poured into control test-tubes. All 24 test-tubes received a suspension of yeast. After two hours Nash measured the amount of carbon dioxide produced in each of the tubes and took periodic measurements for the next six weeks. When he compared the tubes containing the 'held' solutions with the controls, he discov-ered that the solution held by the psychotics had marginally prevented the yeast from growing.[22]

Even deeply buried feelings might have an effect on people we purport to care about. In 1966, Dr Scott Walker of the University of New Mexico School of Medicine conducted a study of alcoholics in the midst of rehabilitation. He divided the group randomly and had members of the Albuquerque Faith Initiative pray for them each day for six months. Half of the participants (some from the treatment group and some from the controls) knew they were being prayed for by family members.

At the end of the six months, Scott discovered that those in both groups whose relatives and friends were praying for them were drinking *more* heavily than the others. Prayer from those who supposedly had the patients' best interests at heart was having the opposite effect.

Scott came up with an interesting interpretation. The across-the-board negative effect of prayer by relatives may reflect their compli-cated, unconscious feelings towards alcoholics. Although consciously they might wish for their loved one to recover, they might actually wish for them to carry on drinking, if the person praying is a fellow drinker and does not wish to lose a drinking buddy. Or, perhaps the boorish, selfish behaviour of an alcoholic has so hurt the relatives that they unconsciously wish for the alcoholic to die.

All these studies are small, but they carry a huge implication: *even your current state of mind carries an intention that has an effect on life around you.* The mind continues affecting its surroundings whether or not we are consciously sending an intention. To think is to affect. When we are consciously attempting to affect someone else with our thoughts, we may want to search our hearts about our true feelings to ensure that we are not sending tainted love.

These studies also raise the possibility that the thoughts that spill out of us at every moment also affect inanimate objects within our reach. Some people have a reputation for having a positive or negative effect on electronic equipment – they are either an 'angel' or a 'gremlin'. One of the fathers of quantum theory, the brilliant theoretical physicist Wolfgang Pauli, was widely known to possess a powerfully negative force field. Whenever he arrived at his laboratory, mechanisms would freeze, collapse or even be set alight.[23]

I am a gremlin of the first order. In those rare moments when I am crashing around in a bad mood, all the computers in our office begin crashing in unison. Once, during a day of extreme agitation, after I had broken my computer and printer at home, I headed off for work and tried to work on a variety of computers around my company's office. One by one, they died in my hands. When one of our laser copier printers also froze the moment I tried to photocopy a page, my team firmly but politely escorted me off the premises.

The late Jacques Benveniste discovered the gremlin effect first hand when he carried out experiments on electromagnetic signalling between cells. From 1991, after his noted 'memory of water' studies, Benveniste understood that the basic signalling between molecules was not chemical but electromagnetic. Within a living cell, molecules communicate, not by chemicals but by electromagnetic signalling at low frequencies, and each molecule has its own signature frequency.[24] Until the end of his life in 2005, Benveniste explored the possibility that these molecular signals could be transferred simply by using an amplifier and electromagnetic coils. He demonstrated that it was possible to effect a molecular reaction without the presence of the molecule in question simply by playing the molecule's unique 'sound'.

One of Benveniste's many experiments with cellular signalling concerned the interruption of the coagulation of plasma, the yellowish medium of the blood. Ordinarily caused by the presence of calcium in the liquid, the clotting capacity of plasma can be precisely controlled by first chemically removing all existing calcium in the plasma, then adding back particular amounts of the mineral. By also adding heparin, an anticoagulant drug, the plasma is prevented from clotting, even in the presence of calcium.

In his study Benveniste would remove calcium from the plasma and add calcium to water, but instead of adding the actual heparin to the calcium water, he simply exposed the water containing calcium to the 'sound' of heparin transmitted through the digitized electromagnetic frequency of heparin that he had discovered. As with all his other experiments, the signature frequency of heparin worked as though the molecules of heparin itself were there: in its presence, the blood was less able to coagulate.

Benveniste had a robot built to carry out this experiment, largely to silence his critics by eliminating the potential bias of human interference. The robot was a box with an arm that moved in three directions, mechanically exposing the water containing calcium to the heparin in several easy steps.

After hundreds of such experiments, Benveniste discovered that it usually worked well except on days that a certain woman – an otherwise experienced scientist – was present. Benveniste suspected that the woman must be emitting some form of waves that were blocking the signals. He developed a means of testing for this, and discovered that the woman emitted powerful, highly coherent electromagnetic fields that appeared to interfere with the communication signalling of his experiment. Somehow, the woman acted as a frequency scrambler. To test this further, he asked the woman to hold a tube of homeopathic granules in her hand for 5 minutes. When he later tested the tube with his equipment, all molecular signalling had been erased.

Since the problem was likely to be electromagnetic, the obvious next step was to protect the machine from EMFs by building a shield. But once the shield was in place, the machine stopped producing good results. Benveniste pondered this development for some days. Perhaps it had to do with positive effects of the environment, and not simply the absence of negative effects. He opened the shield and asked the man who had been in charge of the lab for many years to stand in front of the robot. Immediately, the robot began again to crank out perfect results. As soon as the man left and the shield was put up, the robot no longer produced decent data. This suggested that, just as some people inhibited equipment, others enhanced it. The shield, originally erected to stop negative influences, had blocked positive ones as well.

Benveniste reasoned that the only substance near the robot capable of picking up positive or negative activity was the tube of water, so he asked the head lab technician to hold the tube in his pocket for two hours. He then put the tube into the machine, removed the man from the room and put up the shield. After that, the robot's experiments worked virtually 100 per cent of the time.[25]

These anecdotal stories of the gremlin effect are not so farfetched when you consider the mountains of data generated by the PEAR laboratory, demonstrating that human intention has the ability to make the random output of computers more orderly even when the intention is not conscious or deliberate. Living consciousness might have a major effect on microprocessor technology, which is now exquisitely sensitive. The tiniest disturbances in a quantum process can be highly disruptive. My own gremlin effect appears to be linked to moments of extreme stress or agitation but for some people it may be the very nature of their thought system.

The idea that we can 'charge' an inanimate object with our thoughts is the basis of the dark arts of many native cultures, which infuse effigies and voodoo dolls with negative intention and then use them to target enemies. There is a rich tradition of using effigies, but not much scientific study of them. Dean Radin once designed an experiment to test the effectiveness of voodoo dolls as an instrument of positive intention. He constructed a tiny effigy of a person known to a group of volunteers, who then directed their prayers to the doll. The prayers turned out to have demonstrable effects – an instance of benevolent voodoo.[26]

If we can be unwitting recipients of negative influence, should we take steps to block it or ward it off? Many psychics recommend using visualization to create a mental image of protection, such as imagining yourself in a giant bubble. Marilyn Schlitz and William Braud tested this idea in a variation on their staring studies with 300 volunteers divided into pairs in separate rooms. One member of each pair (the sender) was asked to use a mixture of imagery and self-regulation techniques like relaxation or Autogenic Training to relax or energize themselves. They were then asked to send an intention to reproduce a similar state in their partner (the receiver) which would be recorded with a polygraph pen. Comparisons of the EDA readings of both senders and receivers showed that the senders had an effect – when they were relaxed or activated, so were their receivers.

The receivers were then asked to visualize a variety of images that would act as a psychological 'shield' to block the senders' influences; any image – a shield, a huge concrete wall, a steel fence, pulsating white light – was suitable, so long as it felt protective. These strategies proved highly successful in blocking one of the unwanted influences.[27]

Then, other scientists from the University of Edinburgh attempted to replicate the EDA studies under more rigorous conditions. The senders alternately attempted to calm or activate the receivers, who were to be open to being influenced for half the session and then to 'block' the influence attempt for the other half, by imagining themsleves wrapped up in a 'shielding cocoon' or adopting a stubborn and uncooperative frame of mind. Nevertheless, during the times of attempted influence, the receivers recorded the same EDA readings, regardless of whether they were 'allowing' or blocking it. If anything, there was a slightly larger effect during the blocking sessions. This suggests that ordinary mental strategies of isolating or protecting ourselves may not be enough to successfully resist unwanted influence.[28]

Qigong practitioners undergo lengthy training to learn techniques enabling them to 'disguise' or make their energy fields temporarily 'invisible' in order to ward off unwanted influence. Creating a psychic shield around yourself to prevent a barrage of negative influences – whether from your boss, a well-meaning but interfering professional, that unfriendly neighbour, even the stranger staring at you in the supermarket queue – is likely to require more than an attitude of resistance or a bit of internal imagery.

Larry Dossey once wrote that the most powerful antidote to negative intention was the line in the Lord's Prayer: 'deliver us from evil'. I came across another more ecumenical instance, from the work of Dr John Diamond, who discovered a simple means of grounding yourself against unwelcome influences. Diamond, a psychiatrist and holistic healer, was inspired by George Goodheart, creator of applied kinesiology, which tests the effect of various substances on the body. Goodheart developed the technique of 'muscle testing', now a feature of applied kinesiology. He would ask a patient to stand facing him, with her left arm out, parallel to the floor: he placed his left arm on the patient's shoulder to steady her, and then asked her to resist with all her strength while he pushed on her arm. In most instances, the arm would spring back and resist the force of Goodheart's push. However when Goodheart exposed that person to noxious substances, such as food additives or allergens, the person's left arm would be unable to resist the pressure of Goodheart's push and easily be overcome.

Diamond applied this muscle testing to toxic thoughts. When a person was exposed to any unpleasant thought, the 'indicator muscle' would test weak. Diamond called it 'behavioral kinesiology' and has tested it on thousands of subjects over many years as a means of instantly taking stock of a person's thoughts and most secret desires.[29]

Diamond discovered one thought that could overcome any sort of negative influence, or debilitating idea or situation. He called it a 'homing thought', because it reminded him of his youth in Sydney, Australia, swimming in the surf. Whenever a large wave threatened, he and his friends would dive to the bottom of the water and hold on to the sand with their fingertips. 'We had learned that as soon as we were faced with this situation of stress, we could dive down, grab on to our securing handhold and hang on to our "rock" until the stress passed,' he writes.[30]

The homing thought that each of us can hold on to, Diamond realized, was our ultimate aspiration or purpose in life. He has also referred to it as 'cantillation': each person's special gift or talent that not only gives one a sense of joy but also union with the Absolute. The term 'homing thought' also reminded him of the direction finder that lost aeroplane pilots use to find their way home. The homing thought can act as a homing beacon for everyone, particularly during the most difficult moments. 'It holds us steadfast,' he once wrote, 'on our course.'

Diamond's ideas have not been subjected to scientific scrutiny, but the sheer weight of his anecdotal evidence in using behavioural kinesiology on thousands of patients lends them a certain significance. Whenever we are besieged by the darkest of intentions, we might best protect ourselves when holding on to the thought of what we have been born to do.

CHAPTER ELEVEN

Praying for Yesterday

ON THE EVE OF THE MILLENNIUM, Leonard Leibovici, an Israeli professor of internal medicine in Israel and expert on hospital-acquired infections, conducted a study of healing prayer's effect on nearly 4000 adults who had developed sepsis while in the hospital. He set up a strict protocol, using a random number generator to randomize the participants into two groups, only one of which would be prayed for, and throughout the study maintained impeccable blinding; neither the patients nor the hospital staff knew who was getting treated – or indeed even knew that a study was being carried out. The names of all those in the treatment group were then handed to an individual, who said a short prayer for the well-being and full recovery of the treated group as a whole.

Leibovici was interested in comparing three outcomes between the prayed-for and not-prayed-for groups: the number of deaths in hospital; the overall length of stay in hospital; and the duration of fever. When calculating the results, he was careful to employ several statistical measurements to examine the significance of any differences. As it happened, the group that had been prayed for suffered fewer deaths than the controls (28.1 versus 30.2 per cent), although the difference was not statistically significant. What was scientifically significant, however, were major differences between the prayed-for group and the controls related to the severity of illness and the time it took to heal. Those being prayed for had a far shorter duration of fever and hospital stay and, in general, got better faster than the controls.

The subject of Leibovici's research – the healing effects of prayer – of course was hardly new. But his study offered one novel twist. The patients had been in the hospital between 1990 and 1996. The praying was carried out in 2000 – between 4 and 10 years later.

The study was meant to be a spoof. The *British Medical Journal* had published it in its Christmas 2001 issue,[1] which is generally reserved for light-hearted commentary, next to a reindeer-shaped cluster of rogue cells. But Leibovici was not joking. He was trying to make a serious point in the most graphic way he could. Leibovici had a particular affinity for mathematics and statistics, and used them repeatedly in his reviews and meta-analyses when evaluating particular procedures. He had even come to believe that diseases and the success of treatment could be predicted through mathematical models.[2]

But the scientific method, in his view, was being defiled by its careless application to alternative medicine. Two years before, also in the Christmas issue of the *BMJ*, he had published an article claiming that alternative medicine masquerading as scientific medicine was like a cuckoo chick nestling in a reed warbler's nest.[3] The begging noises of the interloper chick are indistinguishable from its warbler counterparts; indeed, as it grows, the cries of the cuckoo are so loud that they match the noise of eight little warblers. The warbler parents ignore any clues that they have an impostor in their midst and continue to nourish the cuckoo chick – to the detriment, even death, of their own offspring. Leibovici was convinced that alternative medicine could not accommodate the demands of scientific rigour – and that we had no business wasting precious time and resources on the cuckoo in the nest.

But with that article, it seemed that Leibovici was the one wasting his time and breath. Most of his colleagues had missed the point so thoroughly that his only recourse was to *show* them. Two years later, almost to the day, his prayer study appeared in the *BMJ*.

He had intended that the study would illustrate that you simply cannot use the scientific method to explain subjective things like prayer. The problem was that every one had taken the study at face value. Dozens of sceptics derided the study. As one correspondent wrote, if it were possible to violate the arrow of time in this way, it would allow one to go back in time and prevent the Holocaust from happening by murdering Hitler.[4]

In support of Leibovici, many scientists interested in psychical research claimed that the study offered proof that prayer was effective at any point in time: Larry Dossey, who has also written extensively on 'non-local' consciousness and healing,[5] commented that, in a stroke, Leibovici had turned 'conventional notions of time, space, prayer, consciousness and causality' on their heads.[6] Many others commented that Leibovici had been undone by the very meticulousness of his study design. Leibovici's study had used only one supplicant to carry out the prayers and had sent the same prayer at the same time for each patient in the treatment group, so many of those in the alternative medicine camp did not believe the study suffered from some of the same problems in design as the other prayer research. To all the correspondents, Leibovici retorted in the *BMJ* letters section:

> The purpose of the article was to ask the following question: Would you believe in a study that looks methodologically correct but tests something that is completely out of people's frame (or model) of the physical world, for example, retroactive intervention or badly distilled water for asthma?[7]

It was wrong, he was saying, because it *had* to be wrong. It was statistics tied up in a knot and gone berserk. So that his motive would be clear, he added:

> The article has nothing to do with religion. I believe that prayer is a real comfort and help to a believer. I do not believe it should be tested in controlled trials.

Instead, the true purpose was:

> To deny from the beginning that empirical methods can be applied to questions that are completely outside the scientific model of the physical world. Or in a more formal way, if the pre-trial probability is infinitesimally low, the results of the trial will not really change it, and the trial should not be performed.

Although he had intended to use science to prove the absurdity of alternative medicine, he had actually ended up proving to many people that we can pray today to affect something that occurred yesterday. Leibovici appeared to deeply regret his study and refused to discuss it further.[8] Despite all his efforts throughout his career to apply reason and logic to medicine, this was the study that he would be most remembered for – a study that demonstrated, in effect, that we can go back and change the past.

One of the most basic assumptions about intention is that it operates according to a generally accepted sense of cause and effect: the cause must always precede the effect. If A causes B, then A must have happened first. This assumption reflects one of our deepest beliefs, that time is a one-way, forward-moving progression. This assumption is reinforced every moment of our ordinary lives. First we order our coffee, then the waitress delivers it to our table. First we order a book on Amazon, then it arrives in the mail. Indeed, the most tangible evidence of time's arrow is the physical evidence of our own ageing; first we are born, then we grow old and die. Similarly, we believe that the consequence of our intentions can only occur in the future. What we do today cannot affect what happened yesterday.

However, a sizeable body of the scientific evidence about intention violates these basic assumptions about causation. Research has demonstrated clear instances of time-reversed effects, where effect precedes cause. Leibovici's study was unique among prayer research in that it was conducted 'backward in time' – the healing intention was meant to affect events that had already occurred. But to many frontier scientists, this experiment in 'retro-prayer' simply represented a true-to-life instance of the time-displacement effects regularly seen in the laboratory. Indeed, some of the largest effects occur when intention is sent out of strict time sequence.

Studies like Leibovici's offer up the most challenging idea of all: that thoughts can affect other things no matter when the thought is made and, in fact, may work better when they are not subject to a conventional time sequence of causation.

Robert Jahn and Brenda Dunne at PEAR discovered this phenomenon when they investigated time displacement in their REG trials. In some 87,000 of these experiments, volunteers were asked to attempt to mentally influence the 'heads' and 'tails' random output of REGs in a specific direction anywhere from three days to two weeks *after* the machines had run. As a whole, the 'time-displaced' experiments achieved even greater effects than the standard experiments.[9] Jahn and Dunne had deemed these differences non-significant, only because the number of trials carried out in this manner was tiny compared with the rest of their monumental body of evidence. Nevertheless, the very idea that intention could work equally well whether 'backward', 'forward' or in sequence, made Jahn realize that all of our conventional notions of time needed to be discarded.[10] The fact that effects were even larger during the time-displaced studies suggested that thoughts have even greater power when their transmission transcends ordinary time and space.

Retro-causation has been explored in great detail by Dutch physicist Dick Bierman and his colleague Joop Houtkooper of the University of Amsterdam,[11] and later by Helmut Schmidt, an eccentric physicist at Lockheed Martin who created an elegant variation on time-displaced REG remote influence to determine whether someone's intention could change a machine's output after it had been run. He rewired his REG to connect it to an audio device so that it would randomly set off a click that would be audiotaped and heard through a set of headphones by either the left or right ear. He then turned on the machine and tape recorded their output, ensuring that no one, even himself, was listening. After making copies of this master tape (again, with no one listening), he locked the master tape away, to eliminate the possibility of fraud, and gave medical students the copies a day later. The volunteers were asked to listen to the tape and send an intention to have more clicks in their left ears. Schmidt also created control tapes by running the audio device but not asking anyone to attempt to influence the left–right clicks. As expected, the right and left clicks of the controls were distributed more or less evenly.

Once the participants had finished their attempts to influence the tapes, Schmidt had his computer analyse both the student tapes and the master tape that had been hidden away to see if there was any deviation from the typical random pattern. In more than 20,000 trials carried out between 1971 and 1975, Schmidt discovered a significant result: on both the copies and the masters, 55 per cent had more left-hand than right-hand clicks. And both sets of tapes matched perfectly.

Schmidt believed he understood the mechanism for his improbable results. It wasn't that his participants had changed a tape after it had been created; their influence had reached 'back in time' and influenced the machine's output at the moment that it was first recorded.[12] They had changed the output of the machine in the same way they might have if they had been present at the time it was being recorded. They did not *change* the past from what it was; they *influenced the past when it was unfolding as the present* so that it *became* what it was.

Schmidt continually refined the design of his 'retro-PK' studies over 20 years, eventually involving martial arts students, who are trained in mind-control. In one study, he used a radioactive-decay counter to generate a visual display of random numbers. The students sat in front of this visual display, and attempted mentally to influence the numbers in a particular statistical distribution. Once again, he achieved a highly significant result, with odds against it being a chance occurrence of 1000 to 1. Somehow, the intention of the students had reached 'back in time' to affect what occurred in the first place.[13]

Time-displaced intention has also been successfully applied to living things. German parapsychologist Elmar Gruber, of the Institut für Grenzgebiete der Psychologie und Psychohygiene in Freiburg, carried out a series of ingenious experiments examining whether the movement of animals and humans can be influenced after the fact. His first series of tests concerned gerbils running in activity wheels and moving about within a large cage. A special counter kept track of the number of revolutions in the activity wheel. A beam of light in the cage also had a recording device to note whenever the gerbil made contact with it. Similarly, he asked a group of human volunteers to walk around an area across which he had placed a photobeam, which was also attached to a recorder to note every instance that the volunteers ran into it.

Gruber then converted each revolution of the wheel or contact with the photobeam into a clicking sound. Tapes were made of the clicks, which were copied and stored, again to eliminate fraud. Between one and six days later, volunteers were asked to listen to the tapes and attempt to mentally influence the gerbils to run faster than normal, or the people to run into the beam more often than usual. Success would be measured by a greater number of clicks than usual. Gruber carried out each type of trial 20 times, and in each instance, compared the volunteers' tapes with tapes made during sessions when the animals and humans were not subjected to the remote influence. Four of the six batches of trials achieved significant results, and in three of these, the effect size was larger than 0.44.

An effect size is a statistical figure used in scientific research to demonstrate the size of change or outcome. It is arrived at by a number of factors, usually by comparing two groups, one of which has made the change. An effect size under 0.3 is considered small, between 0.3 and 0.6 is medium, and anything above 0.6 is considered large. Aspirin, considered one of the most successful heart attack preventives of modern times, has an effect size of just 0.032, more than 10 times smaller than Gruber's overall effect size. In the case of the activity-wheel gerbil trial, the effect size was a huge 0.7.[14] If his results had concerned a drug, Gruber would have discovered one of the greatest lifesavers of all time.

Gruber carried out six more intriguing experiments. In one study he recorded the number of times that people in a Viennese supermarket crossed a photobeam, and then recorded the number of times a photobeam was crossed by cars passing through various tunnels in Vienna during the rush hour. These again were converted into clicks, and the tapes made of the clicks were stored for one to two months before being played to volunteers, who were asked to influence the speed of the people on foot or in the cars. This time, he decided to include among his group of influencers some people with psychic ability. He also created similar tapes as controls, which were not exposed to remote intention. Once again, when compared with sessions that were not subjected to influence, the results were highly significant; all but one of the automobile–tunnel studies had a significant effect size; in two of the studies, the effect sizes (0.52 and 0.74) were enormous.[15]

Is it possible to retroactively prevent a disease, after it has infected its host and spread? The Chiron Foundation in the Netherlands designed an intriguing study to test this seemingly impossible proposition. A large group of rats was randomly divided in two groups, and one group given a parasitic infection of the blood. The experiment was blinded so that the experimenters themselves did not know which animals were infected and which were controls until after the study was completed. A healer given photographs of the rats after they had been infected with the disease was asked to attempt to prevent the spread of the parasites. Measurements of the blood cells were taken at several intervals after the animals had been infected. The study was carried out three times, each involving a large number of rats. Two achieved a medium (0.47) effect size.[16]

Psychologist William Braud then asked one of the most provocative questions of all: is it possible to 'edit' one's own emotional response to an event? To test this, he designed a batch of studies to test time-displaced influence on nervous activity. He recorded several tracings of the electrodermal activity (EDA) of volunteers, using standard lie-detection equipment – a reasonable gauge of whether a person is calm or agitated. Braud then asked the participants to examine one of their own tracings and to attempt to influence it, by sending an intention either to calm down or activate their own sympathetic nervous system at that earlier point in time. The other tracings of the participants, which were not exposed to mental influence, were to act as controls. Later, when he compared the tracings with controls, he discovered that those tracings that were exposed to the volunteers' own retro-influence were calmer than the controls. Overall, these studies achieved a small, significant effect size (0.37), offering some of the first evidence that human beings might be able to rewrite their own emotional history.[17] Helmut Schmidt successfully employed a similar study design to change his own prerecorded breathing rate, demonstrating that it is possible to retroactively change your own physical state as well.[18]

Dean Radin set up an EDA test similar to Braud's, but added remote distance to a test of retroactive influence. Two months after running the tests, Radin sent copies of the electrodermal readouts to healers in Brazil and asked them to attempt to quiet the readings. After 21 such studies Radin achieved a 0.47 effect size, similar to Braud's.[19]

Radin also tested the possibility that, under certain conditions, a future event can influence an earlier nervous-system response. He made ingenious use of a strange psychological phenomenon called the 'Stroop effect', named after its discoverer, psychologist John Ridley Stroop,[20] originator of a landmark test in cognitive psychology. The Stroop test uses a list of the names of colours (e.g. 'green') printed in different coloured inks. Stroop found that when people are asked to read out the name of a colour as quickly as possible, they take much longer if the name of the colour does not match the colour of the ink used (e.g. if the word 'green' is printed in red ink) than they do if the name and the colour of the ink match (e.g. if the word 'green' is printed in green ink).

Psychologists believe that this phenomenon has to do with the difference in the time it takes the brain to process an image (the colour itself), compared with the time it takes to process a word (the colour name).

Swedish psychologist Holger Klintman devised a variation on the Stroop test. Volunteers were asked first to identify the colour of a rectangle as quickly as they could, then asked whether a colour name matched the colour patch they had just been shown. A large variation occurred in the time it took his volunteers to identify the colour of the rectangle. Klintman discovered that the identification of the rectangle colour was faster when it matched the colour name shown subsequently.[21] The time it took for people to identify the colour of the rectangle seemed to depend on the second task of determining whether the word matched the rectangle colour. Klintman called his effect 'time-reversed interference'. In other words, the later effect influenced the brain's reaction to the first stimulus.

Radin created a modern version of Klintman's study. His participants sat in front of a computer screen and identified the colours of
rectangles that flashed up on the screen as quickly as possible by
typing in their first letter. The image on the screen would then be
replaced by the name of a colour, and the volunteer would then have
to type either 'y' (yes) to indicate that the name of the colour
matched the colour of the rectangle or 'n' (no) to indicate a
mismatch. Radin varied the second part of the design, so that, after
the participant had identified the colour of the rectangle, he or she
would also have to type in the first letter of the actual colour of the
letters of the colour's name. For instance, if the word 'green' flashed
up but was coloured blue, he or she would have to type in 'b'.

In four studies of more than 5000 trials, all four showed a retrocausal effect. A significant correlation was observed in two of the
studies, with a third marginally significant.[22] *Somehow, the time it took
to carry out the second task was affecting the time it took to carry out the
first one.* Radin concluded that his studies offered evidence of a time
displacement in the nervous system. The implications are enormous.
Our thoughts about something can affect our past reaction times.

One scientifically accepted way to examine the overall power of an
effect is to pool the results of all the studies together into what is
called a 'meta-analysis'. Analysed in this manner, 19 of the retro-
influence studies yielded an extraordinary collective result.[23] William
Braud calculated that the overall effect size was 0.32. Although that is
considered a small effect on its own, it represents ten times the effect
size for most prescription drugs, such as the beta-blocker propanolol,
that are recognized as extremely effective.

A different type of analysis of all the best studies of time displacement was carried out in 1996 by Dick Bierman. In statistics, the best
way to judge an effect is to work out how much it deviates from the
mean, or average. One method popular with statisticians is to work
out the chi-square distribution, which entails plotting the square of
each individual score. Any deviation from chance, whether positive

or negative, will show up as a large positive deviation in bold relief. Bierman detected an enormous variance in individual studies, but collectively they produced results whose occurrence by chance alone was an extraordinary 630 billion to one.[24]

One interpretation of the laboratory evidence of retro-influence suggests the unthinkable: intention is capable of reaching back down the time line to influence past events, or emotional or physical responses, at the point when they originally occurred. The central problem of going 'back to the future' and manipulating our own past are the logical knots the mind gets tied up in when considering them. As British philosopher Max Black argued in 1956, if A causes B, but occurs after B, B often precludes A. Therefore, A cannot cause B.

This conundrum was overlooked in the movie *The Terminator*. If the Schwarzenegger cyborg goes back in time and kills Sarah Connor so that she cannot give birth to future rebel John Connor, there would be no future revolution between man and machine. The Terminator no longer has any need to come back in time or, indeed, no longer any purpose for being created.

British philosopher David Wiggins constructed a similar scenario to illustrate the logical problems inherent in the idea of a time machine. Suppose a young man is the grandson of the cruel leader of a fascist movement. He decides to travel back in time to kill his grandfather, to prevent him from taking control. But if he does so, the young man's mother may not be born and he of course would cease to exist.

Nevertheless, physicists no longer consider retro-causation inconsistent with the laws of the universe. More than 100 articles in the scientific literature propose ways in which laws of physics can account for time displacement.[25] Several scientists have proposed that scalar waves, secondary waves in the Zero Point Field, enable people to engineer changes in space-time. These secondary fields, caused by the motion of subatomic particles interacting with the Zero Point Field, are ripples in space-time – waves that can travel faster than the speed of light. Scalar Field waves possess astonishing power: a single unit of energy produced by a laser in such a state would represent a larger output than all the world's power plants combined.[26]

Certain technologies, such as quantum optics, have made use of laser pulses to squeeze the Zero Point Field to such a degree that it creates negative energy.[27] It is well accepted in physics that this negative energy, or exotic matter, is able to bend space-time. Many theoreticians believe that negative energy would allow us to travel through wormholes, travel at warp speed, build time machines and even help human beings to levitate.

When electrons are packed densely together, the density of the spray of virtual particles that are constantly created in the Zero Point Field is increased. These spray densities are organized into electromagnetic waves that flow in two directions, and so may be going 'back and forward' in time.[28]

Physicist Evan Harris Walker first proposed that retro-influence can be explained by quantum physics if we just take account of the observer effect.[29] Walker and later Henry Stapp, an elementary particle physicist at the University of California at Berkeley, who acted as an independent monitor of Helmut Schmidt's final martial arts study, believed that a small tweak in quantum theory, making use of 'non-linear quantum theory', could explain all cases of retro-influence. In a linear system such as current quantum mechanics, the behaviour of a system can be easily described: $2 + 2 = 4$. The system's behaviour is the sum of its parts. In a non-linear system, $2 + 2$ may equal 5 or even 8. The system's behaviour is more than a sum of its parts – by how much more we can't often predict.

In Walker's and then Stapp's view, turning quantum theory into a non-linear system would enable them to include one other element in the equation: the human mind. In Schmidt's martial arts study, the numbers on the visual display remained in their 'potential' state of all possible sets of numbers until they had been observed by the students. At that point, the mental intent of the students and the numbers on the display interacted in a quantum way. According to Stapp, the physical universe exists as a set of 'tendencies' with 'statistical links' between mental events. Even though the tape of the numbers has been generated, they divide into a number of channels of all possible outcomes. When a person looks at the numbers, his brain state will also divide into the same number of channels. His intent will select out a particular channel, and through the numbers 'collapse' the channels into a single state.[30] Human will – our intention – creates the reality, no matter when.

The other possibility is that all information in the universe is available to us at every moment, and time exists as one giant smeared-out present. Braud has speculated that forebodings of the future might be an act of backward time displacement – a future event somehow reaching back in time to influence a present mind. If you simply reversed presentiment and call it backward influence, so that all future mental activity influences the present, you maintain the same model and results as the retro-causation studies. All precognition might be evidence of backward-acting influence;[31] all future decisions may always influence the past.

There is also the possibility that at the most fundamental layer of our existence there is no such thing as sequential time. Pure energy as it exists at the quantum level does not have time or space, but exists as a vast continuum of fluctuating charge. We, in a sense, are time and space. When we bring energy to conscious awareness through the act of perception, we create separate objects that exist in space through a measured continuum. By creating time and space, we create our own separateness and indeed our own time.

According to Bierman, what appears to be retro-causation is simply evidence that the present is contingent upon future potential conditions or outcomes, and that non-locality occurs through time as well as space. In a sense, our future actions, choices and possibilities all help to create our present as it unfolds. According to the view, we are constantly being influenced in our present actions and decisions by our future selves.

This explanation was bolstered by a simple thought experiment carried out by Vlatko Vedral and one of his colleagues at the University of Vienna: Caslav Brukner, a Serb who had managed to leave Yugoslavia during the civil war and, like Vedral, spent time at Zeilinger's Viennese lab.

When Brukner joined Vedral in London during a year-long fellowship at Imperial College, he began thinking about quantum computation, and the fact that it is billions of times faster than classical computing. Once a quantum computer is finally perfected it will enable one to scan every last corner of the Internet in half an hour.[32] Could this enormous advantage in speed have some basis in Bell's inequality, the famous test of non-locality? Bell demonstrated that the remote influence maintained between two quantum subatomic particles, even over vast distances, 'violates' our Newtonian view of separation in space.

Could this same test be used to show when temporal constraints – the limits governing time – are also violated? Brukner enlisted Vedral to design a thought experiment with him. Their experiment rested on a given in science about time: in the evolution of a particle, a measurement taken at a certain point will be utterly independent of a measurement taken later or earlier. In this instance, the 'inequality' of Bell's would refer to the difference between the two measurements when taken at different times.

For their experiment, they no longer needed two particles, and so could utterly eliminate the 'Bob' particle and concentrate on the photon, 'Alice'. The task now was to make theoretical calculations of Alice's polarization at two points of time. If quantum waves behave like a wriggling skipping rope being shaken at one end, the direction in which the rope is pointed is called polarization. To work out their time sequences mathematically, Brukner and Vedral made use of what is called 'Hilbert', or abstract, space.

First they calculated Alice's polarization, then they measured it moments later. When they had finished their calculations of Alice's current position, they went back and measured her earlier polarization again. They discovered that, between two points of time, Bell's inequality indeed had been violated; they got a different measurement of the first polarization the second time around. The very act of measuring Alice at a later time influenced and indeed *changed* how it was polarized earlier.

The implications of their astonishing discovery were not lost on the scientific community. *New Scientist* included their discoveries in a dramatic cover story: 'Quantum entanglement: How the future can influence the past' and concluded:

> quantum mechanics seems to be bending the laws of cause and effect ... entanglement in time puts space and time on an equal footing in quantum theory ... Brukner's result suggests that we might be missing something important in our understanding of how the world works.[33]

For me, Brukner's thought experiment held a significance far greater than a simple theoretical one. It showed that instantaneous cause and effect not only occurs through space but also back and possibly forward through time. It offered the first mathematical proof that the actions of every moment influenced and *changed* those of our past. It may well be that every action we take, every thought we have in the present, alters our entire history.

Even more significantly, his experiment demonstrated the central role of the observer in creating, and indeed changing, reality. Observing had played an integral part in changing the state of the photon's polarization. The very act of measuring an entity at one point of time changed its earlier state. This may mean that every observation of ours changes some earlier state of the physical universe. A deliberate thought to change something in our present could also influence our past. The very act of intention, of making a change in the present, may also affect everything that has led to that moment.

This sort of backward influence resembles the non-local correlations found in the quantum world, as if the connections were always there in some underlying arrangement.[34] It may be that our future already exists in some nebulous state that we actualize in the present. This makes sense since subatomic particles exist in a state of potential until observed or thought about. If consciousness operates at the quantum frequency level, it would naturally reside outside space and time, and we would theoretically have access to information – 'past' and 'future'. If humans are able to influence quantum events, they are also able to affect events or moments other than in the present.

Radin discovered more evidence that our psychokinetic influence is operating 'backwards' in an ingenious study examining the possible underlying mechanism of intention on the random bits of an REG machine. Radin first ran five REG studies involving thousands of trials, then analysed the experiments through a process called a 'Markov chain', which allowed a mathematical analysis of how the REG machine's output changed over time. For this process, he made use of three different models of intention: first, as a forward-time casual influence (the mind 'pushes' the REG in one direction throughout the influence); second, as a precognitive influence (the mind intuits the precise moment to hit the REG in its random fluctuations to produce the intended result by 'looking into the future' and passively 'bringing back' this information to the present); and third, as a true retrocausal influence (the mind first sets the future outcome and applies all the chain of events that will produce it 'backward' in time).

Radin's analysis of the data had one inescapable conclusion: this was not a process running forward in time, in an attempt to hit a particular target, so much as an 'information' flow that had travelled back in time.[35]

But just how much of the past could we change in the sticks-and-stones world of real life? William Braud had pondered this issue at length. He once observed that those moments in the past most open to change might be 'seed' moments when nature has not made up its mind – perhaps the earliest stages of events before they blossomed and grew into something static and unchangeable.[36] These moments were analogous to a sapling that could still be bent and trained before its trunk was too stiff and branches too large; the brain of a child, which is far more open to influence and learning than an adult's; or even a virus, which is far easier to overcome in its infancy.[37] Random events, decisions with equally likely choices, or illness – all probabilistic moments disposed to early influence where human intention could slightly shift the outcome in a certain direction – might comprise the events in our lives most open to retro-influence. Braud referred to them as 'open', or labile, systems – those most open to change.

These systems include many of the workings of living things, which are random processes, much like the quantum systems of random-event generators. Any one of a number of the biological processes in living things requires a cascade of processes, which would be sensitive to the kind of subtle effects on REG machinery observed, say, in the PEAR research.[38]

In Braud's earlier work, he had discovered that remote influence had its greatest effect when there was a strong need for it.[39] The necessity of a particular outcome might be the one quality that moves mountains backward in time.

A clue to the extent of our reach was revealed in Schmidt's discovery of an observer effect in his audio REG experiments that is much like the effect in quantum experiments: it was most important that the person attempting to influence his tapes be the very first listener. If anyone else heard the tape first and listened with focused attention, it was less susceptible to influence later. A few studies even suggest that observation by any sentient being – human or animal – blocks future attempts at time-displaced influence.

Bierman tested this by rigging up a radioactive source to trigger beeps that were delayed for one second and then observed by a final observer. In about half of the events, another pre-observer was given feedback of this quantum event before the final observer witnessed it. In those instances, the pre-observer's observation resulted in a collapse of the superposition state of the quantum event while, in the other half of cases, the final observer 'produced' the collapse.[40]

If this consciousness is the crucial ingredient for 'collapse' to occur, humans – and their ability to 'reduce' reality to limited states – are completely responsible for the idea that time is an arrow in one direction. If our future choice of a particular state is what affects its present 'collapse', the reality may be that our future and present are constantly meeting up with each other.

This accords with what is understood about the observer effect in quantum theory – that the first observation of a quantum entity 'decoheres', or collapses, its pure state of potential into a single state.[41] This rather suggests that, if no one had ever seen Hitler, we might have been able to send an intention to prevent the Holocaust.

Although our understanding of the mechanism is still primitive, the experimental evidence of time reversal is fairly robust. This research portrays life as one giant, smeared-out here and now, and much of it – past, present and future – open to our influence at any moment.

But that hints at the most unsettling idea of all. Once constructed, a thought is lit forever.

CHAPTER TWELVE

The Intention Experiment

SEEING *ACETABULARIA* FOR THE FIRST TIME takes your breath away. The mesmerizing appearance of this common algae of the Caribbean and the Mediterranean has earned a number of poetic nicknames – 'mermaid's wineglass', or 'sombrerillo' in Spanish – and both are fitting. Its slender stem supports a tiny cupped sombrero, like a miniature green umbrella ready to be popped into an underwater tropical cocktail.

For more than 70 years, biology students have marvelled over this tiny plant, not simply for its appearance but for a single bizarre fact of its existence. *Acetabularia* is a freak of nature. From stem to sombrero, the entire plant, measuring up to 5 centimetres, consists of a single cell. Because of this, *Acetabularia*, unlike most living things, can be counted on to behave predictably. The large nucleus of the cell always sits at the rhizoid, the base of the stalk, and divides only when the plant has reached its full height. This uncomplicated structure has helped to unmask biology's greatest mystery: which portion of the plant engineers its ability to reproduce. In the 1930s, the German scientist Joachim Hammerling elected *Acetabularia* as his perfect 'tool organism' to work out the role of a nucleus in plant genetics.

The simplicity of this single-celled organism with its single giant nucleus not only offered up the secrets of the cell in bold relief, it divulged the whole of the building plans of plant life. Working with *Acetabularia* allowed one to sit in stunned witness to the complex morphology of life within the totality of a single cell, large enough to be visible to the naked eye.

Acetabularia also represented a model organism for my first intention experiment. Fritz Popp, who was to perform the experiment with me, believed that if we were going to attempt to carry out my proposal, we needed to begin on the ground floor. For this first experiment, I planned to assemble a small group of volunteers in London, and ask them to use their intention to affect an organism in Popp's lab in Germany. Using *Acetabularia* for our test subject would be analogous to testing a car made of a single moving part. It removes all the variables of a living thing, with its unfathomable number of chemical and energetic processes occurring at every instant.

Humans, for instance, are like a manufacturing plant covering most of the United States. A septillion chemical reactions occur every second in every tablespoon of our cells, tiny explosions that get multiplied by the 50 million million cells of the average human body. In an experiment comparing, say, the growth rates of two sections of the body, it is almost impossible to control for every variable. Growth rates can be altered by food, water, genetics, mood, or even a sudden dip in air temperature.

During our first intention experiment, Popp intended to examine the alteration in the tiny light being emitted from the algae, which was infinitely more subtle than cellular growth rate. Nonetheless, in multicellular living things, even the light that emanates from each cell is subject to a host of influences: the health of the host, the weather and even the activity of the sun.[1] Light can also differ from cell to cell.

With *Acetabularia*, as the light reassuringly derives from its single nucleus, so it is subject to far less fluctuation. With such a primitive organism, Popp explained, it would be possible to demonstrate, with a fair degree of certainty, that any effect, for better or worse, was entirely the result of our remote influence. Only by using such a simple system could we show that our effect was indisputably due to intention and not a dozen other possibilities.

Generally speaking, an increase of photons indicates that a life form is being stressed and a decrease, that its health has improved. If I sent an intention to make the algae healthier, and the photon count went down, it would likely mean that I was having a good effect. If the photon count went up, it was probable that I was, in some way, harming it. Popp has a number of extremely sensitive photocount detectors at his disposal, which can register an intensity of visible light of about 10–17 watts per square centimetre, analogous to the light coming from a candle several kilometres away.[2] This type of ultrasensitive equipment would enable us to record every single hair's breath of difference – even by a single photon – and so determine the extent of our influence.

Popp had reason to be cautious. For 30 years he had faced enormous opposition to his bold assertion that light emanates from living things,[3] and had finally won respect from the physics community. He had set up his international community of likeminded scientists from prestigious centres all over the globe to work on biophoton emissions.[4] By participating in our experiment, he might risk this hard-won reputation and good will. After all, ultimately I was asking this world-renowned physicist to test whether collective positive thinking could change the physical world.

* * *

The results of a number of experiments had suggested that a 'group' consciousness might possibly exist. In their random-event generator experiments, PEAR's Jahn and Dunne found that the influence of pairs of the opposite sex who knew each other had a powerful complementary effect on the machines – roughly three and a half times that of individuals. Two intensively involved people appeared to create six times the 'order' on a random machine. Some couples even produced a 'signature' result, which did not resemble the effects they generated individually.[5]

There was also evidence that a group all intently focused on the same thought registered as a large effect on a REG machine. Roger Nelson, the chief coordinator of the PEAR lab, had come up with the idea of running REG machines continuously during a particularly engaging event, to examine whether the focused attention of a group had any effect on the random output of the machines.

He and Dean Radin developed what they termed 'FieldREG' devices and ran them during a host of events involving the highly focused attention of an audience: intense or euphoric group workshops; religious group rituals; Wagnerian festivals; theatrical presentations; even the Academy Awards. In most instances, their studies showed that multiple minds holding the same intensely felt thought created some kind of deviation from the norm on the equipment.[6]

Nelson had been fascinated by the possibility of a global collective consciousness. In 1997, he decided to place REGs all over the world, have them run continuously and compare their output with moments of global events with the greatest emotional impact. For his programme, which became known as the Global Consciousness Project, Nelson organized a centralized computer program, so that REGs located in 50 places around the globe could pour their continuous stream of random bits of data into one vast central hub through the Internet. Periodically, Nelson and his colleagues, including Dean Radin, studied these outpourings and compared them with the biggest breaking news stories, attempting to root out any sort of statistical connection. Standardized methods and analysis revealed any demonstration of order – a moment when the machine output displayed less randomness than usual – and whether the time that it had been generated corresponded with that of a major world event.

By 2006, they had studied 205 top news events, including the death of the Princess of Wales, the millennium celebrations, the death of John F. Kennedy, Jr, and his wife, and the attempted Clinton impeachment. When Nelson analysed four years' worth of data, a pattern emerged. When people reacted with great joy or horror to a major event, the machines seemed to react as well. Furthermore, the degree of 'order' in the machine's output seemed to match the emotional intensity of the event, particularly those that had been tragic: the greater the horror, the greater the order.[7]

This trend appeared most notable during the events of 9/11. After the twin towers were destroyed, Nelson, Radin and several colleagues studied the data that had poured in from 37 REGs around the world. Individual statistical analyses were performed by Radin, Nelson, computer scientist Richard Shoup of Boundary Institute and Bryan J. Williams, a psychology undergraduate at the University of New Mexico. According to the results of all four analyses, the effect on the machines during the plane crashes was unprecedented. Out of any moment in 2001, the greatest variance in the machines away from randomness took place that day. The results also represented the largest daily average correlation in output between each machine than at any other time in the history of the project.[8] According to the REGs, the world's mind had reacted with a coherent global horror.

Nelson and three independent analysts took apart the data using a variety of statistical methods. Nelson examined his results through the chi-square distribution method, that statistical technique which plots the square of each of the machine's runs, so that any deviation from chance easily shows up. All of the analysts concluded that an enormous increase in 'order' occurred during time frames relating to key moments in the drama (such as, shortly before the first tower was struck), which were likely to be the most intense periods of horror and disbelief.[9] As REGs are designed to control for electrical disturbances, natural electromagnetic fields or increased levels of mobile phone use, the two scientists were able to discard all those possibilities as potential causes.[10]

Furthermore, although activity of the REGs was normal in the days leading up to 9/11, the machines became increasingly correlated a few hours *before* the first tower was hit, as though there had been a mass premonition. This similarity in output continued for two days after the first strike. Williams thought of it as kind of psychic signature, a giant unconscious psychokinetic effect created by 6 billion minds set to react in unified horror.[11] The world had felt a collective shudder several hours before the first plane crash, and every REG machine had heard and duly recorded it.

Although not every analyst agreed with these conclusions,[12] Nelson, Radin and several of their colleagues eventually were able to publish a summary of their findings in the prestigious physics journal *Foundations of Physics Letters*.[13]

Nelson went on to study other events in the wake of 9/11, including the start of the Iraqi war. He compared REG activity with variations in the approval polls of President George W. Bush, to see if he could discover a connection of any kind between the global 'mind' and current American opinions of the president, and whether the REG network reacted most when there were strong feelings of unity and purpose, as the Americans had shared in the wake of 9/11, or when the public mood was polarized, as it had been after the invasion of Iraq and the deposing of Saddam Hussein's regime. After examining 556 separate polls between 1998 and 2004, Nelson's colleague, Peter Bancel, discovered that peaks in variations followed big public changes of opinion of any variety, either for or against the president. Strong emotion, positive or negative – even to presidential decisions – seemed to produce order.

The results of the FieldREG work and the Global Consciousness Project offer several important clues about the nature of group intention. A group mind appears to have a psychokinetic effect on any random microphysical process, even when not focused on the machinery itself. The energy from a collective, intensely felt thought appears to be infectious. There also appears to be a 'dose' effect; the effect on an REG of a load of people thinking the same thought is larger than the effect of a single person. Finally, emotional content or degree of focus is important. The thought has to engulf a group of people in a moment of peak attention, so that every member of the group is thinking the same thought at the same time. A catastrophe is certainly an effective way to snap the mind to attention.

The data from the Global Consciousness Project had one serious limitation. However accurately Nelson had taken the temperature of the world mind, his data simply referred to the effect of mass *attention*. There had been no *intention* to cause change. What would happen if a number of people were not simply attending to something but also trying to affect it in some way? If the focused attention of a group has a physical impact on sensitive equipment, does the signal get stronger when the group is actually trying to change something?

The only systematic study of group intention concerns the so-called Maharishi Effect of Transcendental Meditation™ (TM), the technique first introduced by Maharishi Mahesh Yogi to the West in the 1960s. Over several decades, the TM organization has carried out more than 500 studies of group meditation, with or without intention, to examine whether meditation has a resonance effect on reducing conflict and suffering.

Maharishi Mahesh Yogi postulated that regularly practising TM enabled you to get in touch with a quantum energy field that connects all things. When a group of meditators was large enough, he claimed, their collective meditations caused 'Super Radiance', a term in physics used to describe the coherence of laser light. During

TM, the theory went, the minds of meditators all become tuned to the same frequency, and this coherent frequency begins to order the disordered frequencies around it. Resolution of individual internal conflict leads to resolution of global conflict.

The TM studies claimed to demonstrate effects from two types of meditation. The first was undirected, the simple consequence of a certain percentage of the population meditating. The second resulted from deliberate intention, and required experience and focus; advanced meditators would target a particular area and direct their meditation to help resolve conflict and lower the rate of violence.

The Maharishi's theory rests entirely on the premise that meditation has a threshold effect. If 1 per cent of the population of a particular area practises TM, he claims, or the square root of 1 per cent of the population practises TM–Sidhi, a more advanced type of meditation, conflict of any variety – rates of murders, crime, drug abuse, even traffic accidents – goes down.

Some 22 studies have tested the positive impact of the Maharishi Effect on crime levels. One study of 24 US cities showed that whenever a city reached a point where 1 per cent of the population was carrying out regular TM, the crime rate dropped to 24 per cent. In a follow-up study of 48 cities, those 24 cities with the requisite threshold percentages of meditators (1 per cent of the population) experienced a 22 per cent decrease in crime, and an 89 per cent reduction in the crime trend. In the other 24 cities without the threshold percentage of meditators, crime increased by 2 per cent and the crime trend by 53 per cent.[14]

In 1993, the TM's National Demonstration Project focused on Washington DC during a large upsurge of local violent crime in the first five months of the year. Whenever the local Super Radiance group reached the threshold number of 4000, the rate of violent crime fell and continued to fall, until the end of the experiment. The study was able to demonstrate that the effect had not been due to any other factors, such as police efforts or a special anti-crime campaign. After the group disbanded, the crime rate in the capital rose again.[15]

The TM organization has also targeted global conflict. In 1983 a special TM assembly met in Israel to send intentions through meditation to resolve the Palestinian conflict. During their sessions, they made daily comparisons between the number of meditators working on the project and the state of Arab–Israeli relations. On days with a high number of meditators, fatalities in Lebanon fell by 76 per cent. Their reach apparently extended beyond armed conflict; ordinary violence – local crime, traffic accidents and fires – also all decreased. When analysing their results, the TM group claimed to have controlled for confounding influences such as weather.[16]

TM adepts have also sought to influence the 'misery index' – the sum of inflation and unemployment rates – in the USA and Canada. And indeed, during one concerted effort between 1979 and 1988, the US index fell by 40 per cent and the Canadian index, by 30 per cent.

Another group of adepts sought to influence the monetary growth and crude-materials price indices as well as the American misery index. In this instance, the misery index fell by 36 per cent, and the crude-materials price index fell by 13 per cent. Although the growth rate of the monetary base was affected, it was only by a small margin.[17]

Critics of TM have argued that these effects could easily have been due to other factors – a reduction in the population of young men, say, or better educational programmes in these areas, or even the ebb and flow of the economy – although the TM organization claims to control for such changes.

The problem with these studies, to my mind, is the controversy surrounding the TM organization itself; rumours now abound about data fixing and the infiltration by Maharishi followers into many scientific organizations. Nevertheless, the TM evidence is so abundant and the studies so thorough that it is difficult to dismiss them completely. Furthermore, the studies are regularly published in peer-reviewed scientific journals, and so must meet some level of scientific rigour and critical scrutiny. The sheer bulk of the research argues compellingly that a force outside the understanding of ortho-dox science might be at work.

But even if the results are legitimate, the TM studies, like the REG data, mostly concerns group *attention*. In many instances, the medita-tors are not people who maintain a focused *intention* to change some-thing else.

For three months in the first quarter of 1998, forest fires raged out of control in the Amazonian state of Roraima, 1500 miles northwest of Brasilia, devastating the rainforest. It had not rained for months – an effect blamed on El Niño – and the ordinarily humid rainforest was bone dry, perfect kindling for the fire that had by that time scorched 15 per cent of the state. The rains, usually so copious in this part of Brazil, remained elusive. The UN termed the fire a disaster without precedent on the planet. Water-carrying helicopters and some 1500 firefighters, including recruits from neighbouring Venezuela and Argentina, fought the flames to no avail.

In late March, the weather-modification experts were called in: two Caiapo Indian shamans especially flown to the Yanomami reservation, housing the last of what are believed to be Stone Age tribes. They danced around a bit and prayed, and gathered up a few leaves. Two days later, the heavens opened and it began to pour. Up to 90 per cent of the fire was extinguished.[18]

The Western equivalent of a rain dance is to hope for good weather, and when carried out as a group intention, it may be just as effective. PEAR's Roger Nelson carried out an ingenious little study, after real-izing that the sun shone on graduation day at Princeton for as long as he could remember. Had the desire of the community for a sunny commencement day had a powerful local effect?

He had gathered weather reports for the past 30 years in Princeton and the surrounding areas for the times around graduation day and statistically compared them; Princeton was drier than usual for that time of year, and drier and sunnier than surrounding communities for just that day. If the figures were to be believed, the collective wish for good weather by the community of Princeton may have created some sort of mental umbrella that only stretched to their borders during that single day.[19]

The only other evidence of group mind had been a provocative little double-blind exercise carried out by Dean Radin, who was interested in the claims of Japanese alternative medicine practitioner Masaru Emoto that the structure of water crystals is affected by positive and negative emotions.[20] Emoto claims to have carried out hundreds of tests showing that even a single word of positive intent or negative intent profoundly changes the water's internal organization. The water subjected to the positive intent supposedly develops a beautiful, highly complex crystalline structure when frozen, whereas the structure of water exposed to negative emotions became random, disordered, even grotesque. The most positive results supposedly occur with feelings of love or gratitude.

Radin placed two vials of water in a shielded room in his laboratory at the Institute of Noetic Sciences in Petaluma, California. Meanwhile, a group of 2000 attendees at one of Emoto's conferences in Japan was shown a photo of the vials and asked to send them a prayer of gratitude. Radin then froze the water in those vials as well as samples of control water from the same source that had not been exposed to the prayers, and showed the resulting crystals to a panel of independent volunteers. He had carefully blinded the study so that neither he nor his volunteers had any idea which crystals had been grown from the water samples that had been sent intention. A statistically significant number of the volunteer judges concluded that water sent the positive intentions had formed the more aesthetically pleasing crystalline structure.[21]

Nelson's Global Consciousness Project effects had been an especially intriguing example of the power of mass thought. In a sense, they showed the same effect captured by Tiller's equipment in his laboratory. Intention appeared to be raising order in the ground state of the Zero Point Field. But was there a magic threshold effect, as the Maharishi maintained? And how many people were required to constitute a critical mass? According to the Maharishi's formula – that the square root of 1 per cent of any population practising advanced meditation will have a positive impact – only 1730 advanced American meditators would be required to have a positive influence on the US, and only 8084 to affect the entire world.

Nelson's work with FieldREGs had suggested that the size of the group was not as important as the intensity of focus; any group, however small, exerted an effect so long as the parties were involved in rapturous attention. But how many people did the group need to exert an effect? How intently focused did we need to be? What were the true limits of our influence – if any? It was time for me to find my own answers.

The original plan for our first intention experiment, as Popp saw it, was to gather a group of experienced meditators in London, and to have them send positive intention to the *Acetabularia acetabulum* growing in Popp's IIB laboratory in Neuss, Germany.

I was deflated after we had discussed the likely target. For our first experiment, I had wanted to help heal burn victims, to save the world from global warming. Single-celled organisms weren't exactly my idea of heroics and high drama.

Then I began to research algae, and quickly changed my mind. Vital algae were being killed off as a result of global warming. Scientists have discovered an inexorable rise in ocean temperatures over the past century. For the past 30 years, coral reefs, the centre-piece of the sea's ecosystem, have been vanishing off the earth. When oceans warm, the algae hugging coral reefs get sloughed off, and without this protective layer, the coral reefs themselves die. Some 97 per cent of a certain species of coral have disappeared in the Caribbean alone, and the US government has recently declared Elkhorn and Staghorn coral to be endangered species.

According to the United Nations' Intergovernmental Panel on Climate Change, a body made up of the world's leading climatologists and other scientists, the predicted level of warming – up to 6°C by the end of this century – will bring on a disaster of biblical proportions: a rise of sea levels by nearly 1 metre; unendurable heat in many parts of the world; a vast increase of vector-born diseases; raging floods and storms. A change upward of six degrees may not seem like much until one takes on board that lowering it by the same amount would bring on another Ice Age.

The key to warding off all the fires and floods appeared to be algae. Algae and other plants are the firefighters of our overheated oceans. Scientists are presently engaged in studying sediments from the ocean floor to see how the oceans cope with rising levels of gases. They are especially interested in the reaction of marine plants to global warming, as they are the primary shock absorbers of excess carbon dioxide. Algae provide oxygen and other benefits to plant and animal marine life. Algae offer a little wall of protection to the creatures of the sea from the worse excesses of man.

I reconsidered my resistance to *Acetabularia* as a test subject. Algae might be critical to our survival. The health of most life in the seas depends on these lowly, single-celled creatures, and the seas, like the rainforests, represent the lungs of the earth. As algae goes, so, eventually, do we. Being able to show that mass intention could rescue a sample of algae might demonstrate that our thoughts could combat something as potentially devastating as global warming.

* * *

On 1 March 2006 I travelled to Germany to meet Popp and his colleagues at the IIB laboratory on Museum Island in Hombroich, west of Düsseldorf. The 'island's' innovative architecture had first been built to serve the eccentric needs of a millionaire art collector turned Buddhist, Karl Heinrich Müller, who had nowhere to house his vast collection of painting and sculpture. He purchased 650 acres from the American military, and then converted a NATO missile site into an open-air museum.

Müller's ambitions for the island grew to embrace the possibility of an artists' and writers' community. He commissioned a sculptor turned architect named Erwin Heerich and gave him a free hand. Heerich created enormous futuristic brick structures – galleries, a concert hall, working spaces and even residences – and ingeniously placed them to best advantage against the bleak landscape. Nothing had been wasted; even the disused metal bunkers and rocket silos had been converted into studios and working spaces for famous German artists, writers and musicians, including Thomas Kling the lyricist and Joseph Beuys the sculptor.

Past a 'garden' of buildings of different pastels, the eye alighted on a squat building of interlocking squares on a narrow base, like a giant piece of Lego about to take flight – the new official international site of the IIB. Popp politely accepted the building, when it was first offered to him, but found the open, airy loft, its floor-to-ceiling windows staring out on the vast panorama of Museum Island, completely impractical for his purposes. Before long he set up camp in one of the cramped metal bunkers, left from the *raketenstation*, whose small dark rooms are more compatible with the work of counting living light.

There I met Popp's team of eight, which included Yu Yan, a Chinese physicist, Sophie Cohen, a French chemist, and Eduard Van Wijk, a Dutch psychologist. Most of the cramped rooms contained photomultipliers, large modern boxes attached to computers that count photon emissions. One room housed another smaller room, with a bed and a photomultiplier for human subjects. The pride of place was reserved for a strange homemade contraption of welded metal circles, resembling a David Smith sculpture of scrap metal, which periodically clanged. That, Popp said with pride, was his first photomultiplier, assembled in 1976 by his student, Bernhard Ruth, and still one of the most accurate pieces of equipment in the field. Indeed, he was convinced that it kept improving with age.

When measuring subtle effects, such as the tiny discharges of light from a living thing, it is important to construct a test that will yield a large enough effect to indicate that something has changed. Our experimental design had to be so robust, said Popp, that a positive result could not be dismissed by *advocatus diaboli*, the scientific process of identifying weaknesses in a scientific hypothesis and providing a ready explanation for anomalous effects. Or, as Gary Schwartz had put it, if we heard hoof beats, we first had to eliminate horses before leaping to the conclusion that they belonged to zebras.

In our experimental design, we also had to aim for an 'on off, on off' effect, so that we could isolate any changes as being caused by remote influence. Popp suggested that we have our group send intention intermittently at regular intervals: 10 minutes on, then 10 minutes off, so that we would be 'running' intention a few times every hour. If our experiment worked and intention did have an effect, once we plotted our result on a graph it would create an identifiable, zigzag effect.

Popp acquiesced to including dinoflagellates as well as *Acetabularia*. The light emissions of these fluorescent creatures are extraordinarily responsive to change. As he had seen when they had been placed in shaken water, a change of any sort to which a dinoflagellate is exposed readily shows up as a large shift in emissions of light. I made a further appeal for the use of several subjects. Each would constitute a separate experiment, and then we would have several results to compare. More than one positive finding would be less likely the result of chance. Finally, the scientists agreed. We also added a jade plant, and a human subject whom Eduard felt he could enlist.

As Popp had concluded during his experiment with Dick Blasband, change of any sort is easier to see with something ill that you try to make well, so we needed to stress some of our subjects in some way. The most obvious way to stress a life form is to place it in a hostile medium. Eduard and Sophie decided to pour some vinegar into the medium of the dinoflagellates. We could stress the jade plant by sticking a needle through one of its fleshy leaves. Eduard ultimately decided to stress our human subject with three cups of coffee, but I agreed not to disclose this fact to my meditators, to see if they could pick up any psychic information about her. We decided to leave the *Acetabularia* alone, to test whether our intentions could also affect a healthy organism. To make it simple, our meditators would send intentions for the biophoton emissions of each organism to decrease and for its health and well-being to improve.

The experiment would run at night, between 3 p.m. and 9 p.m. Eduard and Sophie would turn on the equipment, and I would select three half-hour windows within that time frame, unbeknownst to them, to carry out our group intentions. Although it was impossible to conduct a double-blind trial (all of us in London would of course know when we sent our healing intention), we could create 'single-blind' conditions and control for experimenter effects, by ensuring that neither our human subject nor the scientists knew when intention was being sent. I would reveal our schedule to them only after the experiment had taken place.

Our study design was constrained by the equipment. A photomultiplier cannot run with the shutter open continually for six hours, so we decided to turn it on from the hour to the half hour, and give it a rest between the half hour and the hour. I would instruct my meditators to send an intention to all four subjects for two 10-minute sessions during the three time windows I'd chosen. Eduard and Popp planned to look for any qualitative differences in the kind of light being emitted. Any change in the signal or the quantum nature of the photons during the times we were 'running' intention would suggest that change had occurred from an outside influence and that we were having an effect.

I took some photos of our subjects and the scientists. Before leaving, I stole a last look at the *Acetabularia*, growing in small pots in a converted, darkened refrigerator, and the dinoflagellates, which resembled tiny green specks in the water – tiny participants about to be stressed, and possibly sacrificed, in the name of science.

A few weeks later, Eduard found a human volunteer in one of his Dutch colleagues, Annemarie Durr,[22] a laser biologist and a meditator of long standing. Although rather sceptical of our plan, she was happy to be our first subject. Her agreement to participate was a particularly generous gesture, as it would entail sitting still on a bed in a pitch black room for six hours.

At one of our conferences in mid-March, I asked for volunteers to participate in a first intention experiment from those among our audience who were experienced meditators. I prepared a PowerPoint presentation to brief them on the subjects of our experiment and the experimental protocol, and to reinforce my verbal presentation, and set the day for 28 March at 5:30 p.m., at a university lecture room I had hired for the evening.

That night, there was such a fierce hailstorm when my colleague Nicolette Vuvan and I left our office for the train to central London that we had to take momentary shelter in a doorway. We were half soaked after battling through a torrent of rain, but I was thrilled with the atmospheric conditions – a dark, stormy night would only aid our activities. Weather this wild often results from geomagnetic or atmospheric disturbance, which I knew enhances psychokinetic effects. When I checked with America's National Oceanic and Atmospheric

Administration's website later that evening, I discovered that they noted 'unsettled' conditions for the evening, with a fair degree of geomagnetic activity and minor to major storms in space.

Despite the weather, 16 volunteers showed up. I asked them to fill in a collection of forms, which included personal information plus several psychological tests used by Gary Schwartz and Stanley Krippner, including the Arizona Outcome Integrative Scale and the Hartmann Boundary Questionnaire test, to test psychic ability. I wanted as much data as possible in order to gauge whether their state of mind, psychic talent, or health status would have any bearing on our results.

I soon discovered that my volunteers were ideal candidates for an intention experiment. According to the forms they'd filled out for me, they'd meditated for an average of 14 years, and their scores on the psychological tests I'd given them showed that, as a group, they had very thin boundaries, tended toward a highly positive outlook, enjoyed excellent mental, emotional and physical health, and evidenced powerful emotions.

I explained the experiment, offered photographs and details about our four subjects, and then went over the protocol. We would be sending our intentions from 6 p.m. to 8:30 p.m. at every hour on the hour to 10 minutes past and from 20 past until the half hour. In between those times we would rest, chat and fill in the forms.

We began at 6 on the dot. As William Tiller had done in his black box experiments, I displayed the intentions in writing on the computer screen as I read them out loud so that all the meditators would be sending exactly the same thought during each meditation. I led the meditation, directed our focus to each target subject, showing its image on screen, and read aloud the sentence that sent our intention to lower the subject's biophoton emissions and increase its state of health and well-being.

The shared energy immediately felt tangible and increased in power as the evening carried on. Michael, one of our group members, suggested that we call our algae 'Dino' and 'Tabu', to establish some relationship with these little organisms. Although no one had any prior experience in telepathy, some participants began to pick up information about our subjects, notably Annemarie. Several meditators were convinced that she was an amateur singer, and had a recurrent problem with her throat. Isabel thought she might be suffering from gut problems or something gynaecological. Michael, who was German, kept thinking of a term *Im schutz der dunkelheit* ('under protection of darkness'), and interpreted it to mean that she was wrapped up in a blanket. Amy said she received a mental image of Annemarie wrapped up in a luxuriously soft blanket on a hard surface and at times asleep. She was also convinced that she had eaten something disagreeable and that her stomach was upset.

Many meditators felt a connection to the jade plant and 'Tabu', and Peter had a strong sense that *Acetabularia* was responding most to the intentions – but with few exceptions the group had the most difficulty establishing any connection with 'Dino', and this difficulty increased to the point where most felt no connection at all by the final session.

All of us were infused with a strong sense of purpose and momentarily lost a sense of our individual identities. By the end of the evening, I had cast out my own doubts about the study and the niggling thought that what we were trying to do was faintly ludicrous. Even though we were not healers, we had all felt as if a healing of sorts had occurred. Whatever had happened in there, I thought, heading back out into the stormy night, I grew certain we'd had some kind of effect.

Several days later, I sent Popp our meditation schedule so that his team could compile our results. I also spoke with Annemarie. Some of our extrasensory impressions had been correct. It was true she sang as a hobby and periodically suffered with a blocked throat. Although she ordinarily did not especially suffer problems in the gut, she had that night because the three cups of coffee Eduard had asked her to drink upset her stomach. Yet even though coffee late in the afternoon usually agitated her and caused insomnia, on the night of our experiment, she drifted off at various points throughout the six hours of the experiment and slept easily that night. She described tingling bodily sensations she had felt periodically through the evening, and the times of their occurrence corresponded with the first and third sessions that we had been 'running' intention. Nevertheless, we had also picked up some 'noise': she was not a vegetarian and never listened to or had sung Vivaldi, as a couple of meditators had felt.

When analysing the data, Eduard studied not only the intensity of light but also its deviation from symmetry: normal emissions from a living thing, when plotted on a graph as a bell curve, are perfectly symmetrical. He also looked at any deviations in the kurtosis, or the customary 'peakedness', of the distribution. High kurtosis means a bell curve that is high around the middle, or mean. Again, when emissions are plotted on a graph, the normal peak distribution is 0 – the highs and lows cancel each other out. After examining our 12 block periods – the six times we sent intention and the six periods of rest – he found no change in light intensity. But he did find large changes in the skewness, showing a lack of the customary symmetry (from 1.124 to 0.922), and kurtosis (from 2.403 to 1.581) of the emissions. Something in the light was profoundly altered.

Eduard was excited by the results. They exactly matched those he had observed during his study of healers, when he had tested whether the act of healing has a 'scatter effect' on any other living things in the environment where the healing takes place. In the study, when he had placed some algae with a photon counter in the presence of a healer and his patients and measured the photons of the algae during 36 healings, he had been surprised to discover that the photon count distributions of the algae had 'remarkable' alterations during the healing rituals. Large shifts in the cyclical components of the emissions had occurred. His tiny study had suggested that healing caused a shift in the light emissions of everything in its path.[23]

Now he had discovered the same effect when simple intention was sent by ordinary people from 300 miles away.

On 12 April, Fritz Popp sent me data on the algae, the dinoflagellates and the jade plant. Although a first glance at the numbers had convinced him we had had no effect, he changed his mind once he performed his calculations. Ordinarily, any stressed living thing will begin to accustom itself to the stress, and its light emissions, although initially large, will naturally begin to decrease as the organism gets used to its new circumstances. Consequently, in order to work out a true demonstration of the effect of change, Popp had to control for this phenomenon. He worked out mathematically a means of starting from zero, so that any deviation from normal behaviour would readily show up. In this way, he would then be able to determine whether any additional change represented an increase or a decrease in the number of biophoton emissions. The number of emissions he then plotted on his graph reflected any excess increase or decrease from the norm.

In all three instances, our subjects registered a significant decrease in biophotons during the meditation sessions, compared with the control periods. The dinoflagellates had been killed by the acid, in the end (one possible reason why they had been so difficult for our meditators to detect). Nevertheless, Popp said, their response (a lowering of emissions by nearly 140,000) was significantly different from the normal emissions of a dying organism. Among the survivors, the *Acetabularia*, the healthy subject, had evidenced a larger effect than the jade plant, perhaps because it was not overcoming a stress (544 emissions lower than normal), whereas with the jade plant (which had 65.5 emissions lower than normal), the stress (the pin) remained in the leaf during the experiment.

He plotted the results on a graph, marking out the portions in red that represented the times of our healing intentions and emailed them to me. We had indeed produced a 'zigzag' effect. During meditation, Popp wrote in his report, 'there is a clear preference of dropping down reactions rather than going up', which tracked the times of our intentions. With the *Acetabularia*, we had had an overall decrease over the norm of 573 emissions, and an increase of only 29.

Our little meditation effort had created a major healing effect, a significant decrease in living light. Not only that, but the effect from all that distance was similar to the effect by an experienced healer when healing in the same room. The intention of our group had created the same light as a healer's.

In many ways, it was a crude first effort. We had, after all, tested four subjects, some stressed and some not, and one had died. We had made use of control periods, but not control subjects. Both Eduard and Popp cautioned me not to take too much notice of it: 'We have to be sure that these changes in kurtosis and skewness are real. That means that we have to repeat the experiments a couple of times,' said Eduard. 'Despite the right tendency of the results,' wrote Popp, 'I do not dare to state that it is proof.'

But, despite these caveats, the fact was that we had recorded a significant effect. In the end, achieving a positive result didn't really surprise me. For more than 30 years Popp, Schlitz, Schwartz and all of their fellow scientists have been amassing unimpeachable evidence in other experiments that has stretched credulity. Frontier research into the nature of human consciousness has upended everything that we have hitherto considered scientific certainty about our world. These discoveries offer convincing evidence that all matter in the universe exists in a web of connection and constant influence, which often overrides many of the laws of the universe that we used to believe held ultimate sovereignty.

The significance of these findings extends far beyond a validation of extrasensory power or parapsychology. They threaten to demolish the entire edifice of present-day science. The discoveries of Tom Rosenbaum, Sai Ghosh and Anton Zeilinger that quantum effects occur in the world of the tangible could signal an end to the divide in modern physics between the laws of the large and the laws of the quantum particle, and the beginning of a single rule book defining all of life.

Our definition of the physical universe as a collection of isolated objects, our definition of ourselves as just another of those objects, even our most basic understanding of time and space, will have to be recast. At least 40 top scientists in academic centres of research around the world have demonstrated that an information transfer constantly carries on between living things, and that thought forms are simply another aspect of transmitted energy. Hundreds of others have offered plausible theories embracing even the most counter-intuitive effects, such as time-displaced influence, as now consistent with the laws of physics.

We can no longer view ourselves as isolated from our environment and our thoughts the private, self-contained workings of an individual brain. Dozens of scientists have produced thousands of papers in the scientific literature offering sound evidence that thoughts are capable of profoundly affecting all aspects of our lives. As observers and creators, we are constantly remaking our world at every instant. Every thought we have, every judgement we hold, however unconscious, is having an effect. With every moment that it notices, the conscious mind is sending an intention.

These revelations not only force us to rethink what it is to be human, but also how to relate. We may have to reconsider the effect of everything that we think, whether we vocalize it or not. Our relationship with the world carries on, even in our silence.

We must also recognize that these ideas are no longer the ruminations of a few eccentric individuals. The power of thought underpins many well-accepted disciplines in every reach of life, from orthodox and alternative medicine to competitive sport. Modern medicine must fully appreciate the central role of intention in healing. Medical scientists often speak of the 'placebo effect' as an annoying impediment to the proof of the efficacy of a chemical agent. It is time that we understood and made full use of the power of the placebo.

Repeatedly, the mind has proved to be a far more powerful healer than the greatest of breakthrough drugs.

We will have to reframe our understanding of our own biology in more miraculous terms. We are only beginning to understand the vast and untapped human potential at our disposal: the human being's extraordinary capacity to influence the world. This potential is every person's birthright, not simply that of the gifted master. Our thoughts may be an inexhaustible and simple resource that can be called upon to focus our lives, heal our illnesses, clean up our cities and improve the planet. We may have the power as communities to improve the quality of our air and water, our crime and accident statistics, the educational levels of our children. One well-directed thought may be a gentle but effective way for every man and woman on the street to take matters of global interest into their own hands.

This knowledge may give us back a sense of individual and collective power, which has been wrested from us, largely by the current world view espoused by modern science, which portrays an indifferent universe populated by things that are separate and unengaged. Indeed, an understanding of the power of conscious thought may also bring science closer to religion by offering scientific proof of the intuitive understanding, held by most of us, that to be alive is to be far more than an assemblage of chemicals and electrical signalling.

We must open our minds to the wisdom of many native traditions, which hold an intuitive understanding of intention. Virtually all of these cultures describe a unified energy field not unlike the Zero Point Field, holding everything in the universe in its invisible web. These other cultures understand our place in a hierarchy of energy and the value of choosing time and place with care. The modern science of remote influence has finally offered proof of ancient intuitive beliefs about manifestation, healing and the power of thoughts. We would do well to appreciate, as these traditional cultures do, that every thought is sacred, with the power to take physical form.

Both modern science and ancient practices can teach us how to use our extraordinary power of intention. If we could learn how to direct our potential for influence in a positive manner, we could improve every aspect of our world. Medicine, healing, education, even our interaction with our technology, would benefit from a greater comprehension of the mind's inextricable involvement in its world. If we begin to grasp the remarkable power of human consciousness, we will advance our understanding of ourselves as human beings in all our complexity.

But there are still many more questions to ask about the nature of intention. Frontier science is the art of inquiring about the impossible. All of our major achievements in history have resulted from asking an outrageous question. What if stones fall from the sky? What if giant metal objects could overcome gravity? What if there is no end of the earth to sail off? What if time was not absolute, but depends on where you are? All of the discoveries about intention and remote influence have similarly proceeded from asking a seemingly absurd question: what if our thoughts could affect the things around us?

True science, unafraid to explore the dark passages of our ignorance, always begins with an unpopular question, even if there is no prospect of an immediate answer – even if the answer threatens to overturn every last one of our cherished beliefs. The scientists engaged in consciousness research must constantly put forward unpopular questions about the nature of the mind and the extent of its reach. In our group experiments, we will be asking the most impossible question of all: what if a group thought could heal a remote target? It is a little like asking, what if a thought could heal the world? It is an outlandish question, but the most important part of scientific investigation is just the simple willingness to ask the question. As Bob Barth of the Office of Prayer Research commented, when asked whether prayer research should continue in the wake of the Benson STEP study: 'We can't find the answers if we don't keep asking the questions.' That is how we will begin our own experiments – unafraid to ask the question, whatever the answer.

PART FOUR

The Experiments

Miracles do not happen in contradiction
to nature, but only in contradiction to
that which is known in nature.
St Augustine

CHAPTER THIRTEEN

The Intention Exercises

UP UNTIL THIS POINT, *The Intention Experiment* has been concerned with the scientific evidence of the power of intention. What has not been tested is the extent of this power in the cut and thrust of ordinary life. An inordinate number of books have been written about the power of the human being to manifest his or her reality, and, while they have served up many intuitive truths, they offer little in the way of scientific evidence.

Exactly how much power do we possess to shape and mould our daily lives? What can we use this for, individually and collectively? How much power do we possess to heal ourselves, to live lives of greater happiness and purpose?

This is where I would like to enlist your help. Determining the further practical applications of the power of thought is the purpose of the next portion of this book – the part that involves you as a partner in the research.

Although the power of intention is such that any sort of focused will may have some effect, the scientific evidence suggests that you will be a more effective 'intender' if you become more 'coherent', in the scientific sense of the term. To do so to greatest effect, or so the scientific evidence suggests, you will need to choose the right time and place, quiet your mind, learn how to focus, entrain yourself with the object of your intention, visualize and mentally rehearse. Believing that the experiment will work is also essential.

Most of us operate with very little in the way of mental coherence. We walk around immersed in a riot of fragmentary and discordant thought. You will become more coherent simply by learning to shut down that useless internal chatter, which always focuses on the past or the future, never the present. In time, you will become adept at quietening down your mind and 'powering up', much as joggers train their muscles, and each day find that they can perform a little better than the day before.

The following exercises are designed to help you to become more coherent and so more effective in using intention in your life and in our group intention experiments. These have been extrapolated from what has appeared to work best in the scientific laboratory.

Think of intentions in terms of grand and smaller schemes. Take the grand schemes in stages, so that you send out intentions in steps towards achieving the grand scheme. Also start with modest goals – something realizable within a reasonable timeframe. If you are 40 pounds overweight and your goal is to be a size 8 next week, that is not a realistic timeframe. Nevertheless, keep the grand scheme in mind and build towards it as you gain experience. It is also important to overcome your natural scepticism. The idea that your thought can affect physical reality may not fit your current world paradigm, but nor would the concept of gravity if you were living in the Middle Ages.

Choose Your Intention Space

A number of scientific studies suggest that conditioning your space magnifies the effectiveness of your intentions. Choose a place to carry out your intentions that feels comfortable. Clear away extraneous items and make it personal or appealing, with cushions or comfortable furniture, so that whenever you spend time there you will find it an enjoyable refuge, a place where you can sit quietly and meditate. Use candles, soft lights and incense, if you prefer.

Some people find it helpful to create an 'altar' of sorts, as a focal point, with objects or photographs that you find inspirational or particularly meaningful. Even if you are not at home, you may find that you will naturally 'enter' your intention space by visualizing it whenever you want to send an intention.

Unless you live in the mountains and can open your windows to clean mountain air, you also may want to install an ionizer in your space to increase the number of negative ions in your environment.

The half-life of ions – which is related to the amount of time that ions maintain their effective radiation – depends on the amount of pollutants in the air. The cleaner the air, the longer the half-life of small ions, if there is a source of ionization (e.g. running water) present. The best levels of ions are:

- in the uninhabited country, away from industrialized areas;
- near running water, whether a shower or a waterfall;
- in natural habitats;
- in clear sunshine – a natural ionizer;
- after storms;
- in the mountains.

The worst are:

- in enclosed spaces with a gathering of a number of people;
- near television sets and other such electrical appliances, which can give off electric emissions up to 11,000 volts, exposing anything immediately in range to positive charge;
- in cities;

- near industrial sources;
- in smog, fog, dust or haze.

As a rule of thumb, the lower the visibility, the lower the ion concentration. Low visibility is due to the presence of a great number of large particles, which air ions readily latch on to. For those among us who are city dwellers, placing plants and some source of water, like an indoor desk fountain, will help to improve ion levels in intention spaces. Keep your space free of electrical gadgets and computers.

Power Up

In order to 'power up' to peak intensity, you must first slow your brain waves down to a meditative, or 'alpha', state of light meditation or dreaming – when the brain emits frequencies (measured on an EEG machine) of 8–13 hertz (cycles per second).

Sit in a comfortable position. Many people like to sit upright in a hard-backed chair, with their hands placed on their knees. You may also sit on the floor cross-legged. Begin breathing slowly and rhythmically in through the nose and out through the mouth (slowly blow all the air out), so that your in-breath is the same length as your out-breath. Allow the belly to relax so that it slightly protrudes, then pull it back slowly as if you were trying to get it to touch your back. This will ensure that you are breathing through your diaphragm.

Repeat this every 15 seconds, but ensure that you are not over-exerting or straining. Carry on for 3 minutes and then keep observing it. Work up to 5 or 10 minutes. Begin to focus your attention just on the breath. Practise this repeatedly, as it will form the basis of your meditative practices.

To enter an alpha state, the most important feature, as any Buddhist understands, is to still the mind. Of course, just thinking about nothing is often virtually impossible.

After entering the state by concentrating on the breath or focusing on a single object, most meditation schools recommend some sort of 'anchor', enabling you to keep your chattering mind quiet, so that you are allowed to be more receptive to intuitive information. The usual anchors include focusing on:

- the body and its functions, or the breath;
- your thoughts, as though they are floating by on a flying carpet, so that they are not 'you';
- a mantra, such as used in Transcendental Meditation, is usually a 'word' such as OM ('The Field' in Buddhism), AH (the universal truth of life) or HUM (the physical manifestation of the truth – the universe itself). In the early 1970s, many practitioners of TM were given the mantra AH-OM;
- numbers, through silent repetitive counting, either backwards or forwards;
- music – usually something repetitive, such as Bach or chanting;
- a single tone, such as that produced by an Australian didgeridoo;
- a drum or rattle, the repetitive sounds of which have been used by many traditional cultures to still the mind;
- prayer, as with a rosary, since the repetitive sounds still the mind.

Practise until you can comfortably focus on your 'anchor' for 20 minutes or more.

Peak Intensity

Powering up involves developing the ability to attend with peak intensity, moment by moment. One of the surest ways to develop this is to practise the ancient art of mindfulness, espoused as long ago as 1000 BC by Shakyamuni Buddha, founder of modern Buddhism. It is a discipline whereby you maintain clear, moment-to-moment awareness of what is happening internally and externally, rather than colouring your interpretation with your emotions or being engaged 'elsewhere', deep in thought.

More than just concentration, mindfulness requires that you police the focus of your concentration and maintain that concentration in the present. With practice, you will be able to silence the constant inner chatter of your mind and concentrate on your sensory experiences, no matter how mundane – whether it is eating a meal, hugging your child, noticing some pain you are experiencing or just picking some lint off your sweater. It is like being a benevolent parent to your mind – selecting what it will focus on and leading it back when it strays.

In time, mindfulness meditation will also heighten your visual perceptions and prevent you from becoming numb to your everyday experience.

One of the difficulties in incorporating mindfulness into ordinary activity is that it is usually taught at retreats, where participants have the luxury of meditating for hours a day or practising mindfulness by engaging in activities, as it were, in 'slow motion'. Nevertheless, there are ways to adapt many traditional practices for use in your intention meditation.

Once you have achieved your 'alpha state', quietly observe whatever manifests in your mind and body as precisely as you can. Be present and attentive to what is, rather than what your emotions tell you, what you wish were the case, or only what is most pleasant. Do not suppress or banish any negative thoughts, if they are true. One good means of harnessing your mind to the present is to 'come into your body' and feel your body posture.

It is vital that you distinguish mindfulness from mere concentration. The most important distinction is a lack of judgement or reference point about the experience. You attend to every moment in the present without colouring it with preference for the pleasant or distaste for the unpleasant, or even identifying the experience as something happening to you. There is, in short, no 'better' or 'worse'.

- Be aware of all the smells, textures, colours and sensual feelings you are experiencing. What does the room smell like? What taste is in your mouth? What does your seat feel like?
- Be mindful of what is happening internally and externally. Whenever you catch yourself judging what you see, think to yourself, 'I am thinking', and return to observing with simple attention.
- Cultivate the art of simple listening to all sounds in your room: the rumble of a pipe, the honking of a horn, the barking of a dog, a plane flying overhead. Accept all sounds — the noise, chaos or stillness — without judgement.
- Notice other sensations in the room: the 'colour' of the day, the light in the room, any movement carrying on in front of you, the sensations of quiet.
- Try not to try. Work on eliminating your expectations or striving for (and anxiety over) certain results.
- Accept all that happens without judgement. This means putting away all opinions and interpretations of what goes on. Catch yourself clinging to certain views, thoughts, opinions and preferences, and rejecting others. Accept your own feelings and experiences, even the unpleasant ones.
- Try never to rush. If you must rush, be present in the rushing. Feel what it feels like.

Developing Mindfulness in Your Daily Life

Even when you are not using intention, the evidence suggests that you will mould your brain to become better at it if you develop mindfulness in your daily life. Psychologist Dr Charles Tart, one of the world's experts on altered states of consciousness, has a number of suggestions of ways to do so:[1]

- Take periodic breaks during the day in which you have quiet time to be mindful of what is happening internally and externally.
- Whenever you feel your concentration flitting away in your daily activities, sense your breath – it will help to ground you.
- Be mindful of the most mundane of activities, such as brushing your teeth or shaving.
- Start with a small exercise, such as fetching your coat and walking, in which you stay focused completely on what you are doing.
- Engage in mental noting, in which you label an ongoing activity, for example 'I'm putting on my coat', 'opening the door', 'tying my shoes'.
- Use mindfulness in every ordinary situation. When you are preparing dinner or even doing your teeth, be aware of all the smells, textures, colours and sensual feelings you are experiencing.
- Learn to really look at your partner and your children, your pets, your friends and work colleagues. Observe them closely during every activity – every part of them without judgement.
- During some activity, such as breakfast, ask your children to be mindful (without speaking) of every aspect of it. Concentrate on the taste of your food. Look closely at the texture and the colours of it. How does the cereal crunch? How does their juice feel as it cascades down their throats? Become aware of the smells and sounds around you. While you are watching all this, how are the different parts of your body feeling?
- Listen to what your life sounds like – the myriad noises surrounding you every day. When someone speaks to you, listen to the sound of his or her voice as well as the words. Do not think of a reply until he or she has stopped speaking.
- Practise mindfulness in every activity: walking down the street, driving home, in the garden.

- If you are practising these exercises and you happen to bump into someone, do not enter into conversation. Just greet the person, shake hands and stay in the present moment.
- Use mindfulness when you are extremely busy or under a tight deadline. Observe what it is like to hurry or to be under the gun and what happens when you do. How does it affect your equilibrium? Be an observer of yourself in that situation. Can you stay in your body while you are working hard?
- Practise mindfulness while you are standing in line. Experience the feeling of waiting itself, rather than focusing on what you are waiting for. Be aware of your physical movements and your thoughts.
- Do not think about or try to work out your problems. Just deal with whatever daily problem solving is immediately in front of you.

Merging with the 'Other'

Research shows that touch or even focus on the heart or compassionate feelings for the other is a powerful means of causing brain-wave entrainment between people. When two people touch while focusing loving thoughts on their hearts, the 'coherent' heart rhythms of one can entrain the brain of the other.[2]

Before you set your intention, it may be important to form an empathetic connection with the object of your intention.

Establish connection beforehand by the following techniques:

- First send your intention to someone with whom you already have a strong bond – a partner, a child, a sibling, a dear friend.
- With someone you do not know, exchange an object or photograph.
- Get to know the person. Go for a walk with them or meet them first.

- Spend half an hour meditating together first.
- Ask the person to be open to receiving your intention when you are sending it.
- If you are sending an intention to something non-human or inanimate, you can also establish some connection. Find out all you can about the object of your intentions, whether a plant, an animal or an inanimate object. Have it near you for a period before sending your intention. It goes without saying that you should be nice to it – even if 'it' is your computer or photocopier.

Be Compassionate

Use the following methods to encourage a sense of universal compassion during your intention session:

- Focus your attention to your heart, as though you are sending light to it. Observe the light spreading from your heart to the rest of your body. Send a loving thought to yourself, such as 'May I be well and free from suffering.'
- On the out breath, imagine a white light radiating outward from your heart. As you do, think: 'I appreciate the kindnesses and love of all living creatures. May all others be well.' As Buddhists recommend, first think of all those you love, then your good friends. Move on to acquaintances and finally to those people you actively dislike. For each stage, think: 'May they be well and free from suffering.'
- Concentrate on the kindness and compassion of all living things and their contribution to your well-being. Finally, send your message of compassion to all people and living things on earth.
- Practise switching roles with some of your loved ones. Imagine what it is like to be your partner or spouse, your parent, your child. Get inside their shoes and imagine what it

would be like to see the world through their eyes, with their hopes and fears and dreams. Think how you would respond.

- Jerome Stone quotes Sogyal Rinpoche, author of *The Tibetan Book of Living and Dying*,[3] who suggests that we open our hearts every day to the suffering around us, with beggars who pass us by, with the poverty, tragedy and grief we see on our television sets:

> Don't waste the love and grief it arouses; the moment you feel compassion welling up in you, don't brush it aside, don't shrug it off and try quickly to return to 'normal', don't be afraid of the feeling or embarrassed by it, allow yourself to be distracted from it or let it run aground in apathy. Be vulnerable; use that quick, bright up rush of compassion; focus on it, go deep into your heart and meditate on it, develop it, enhance and deepen it. By doing this you will realize how blind you have been to suffering ...[4]

- During your intention, if you are sending healing to someone, first try to put yourself in his situation. Imagine what it is like to be him and to be faced with his current crisis. Try to feel and have empathy for your receiver's suffering. Ask yourself how you would feel if you were suffering in this manner and how you would most want to be healed.
- Now, direct your loving thoughts to the object of your intention. If he or she is present, hold his or her hand.

Stating Your Intention

In your meditative state, state your clear intention. Although many people use the construction 'have always been' – 'I have always been healthy' – I prefer the present tense – of sending your intention to its 'endpoint' *as a wish that has already been achieved*. For instance, if you are trying to heal back pain, you can say, 'My lower back and sacrum are free of all pain and now move easily and fluidly.' Remember to frame your intention as a positive statement; rather than 'I will not have side effects', say, 'I will be free of side effects'.

Be Specific

Specific intentions seem to work best. Be sure to make your intentions highly specific and directed – and the more detailed, the better. If you are trying to heal the fourth finger of your child's left hand, specify that finger and, if possible, the problem with it.

State your entire intention, and include what it is you would like to change, to whom, when and where. Use the following as a checklist (as news reporters do) to ensure you have covered every specific: who, what, when, where, why and how. It may help if you draw a picture of it, or create a collage from photos or magazine pictures. Place this somewhere that you can look at often.

The Mental Dry Run

As with elite athletes, the best way to send an intention is to visualize the outcome you desire with all your five senses in real time. Visualization, or guided imagery, involves using images and/or internal messages to obtain a desired goal. It can be used for any desired outcome – to change or improve your living situation, job, relationships, physical condition or health, state of mind (from negative to positive), outlook on life or even a specific aspect of yourself, including your personality. It can also be used to send intentions to someone else. Self-guided imagery is a little like self-hypnosis.

Plan a mental image of the outcome of your intention well ahead of time. When carrying out visualization, many people believe that you must 'see' the exact image clearly in your mind's eye. But for an intention it isn't necessary to have a sharp internal image or, indeed, any image at all. It is enough to just think about an intention, without a mental picture, and simply to create an impression, a feeling or a thought. Some of us think in images, others through words, still others through sounds, touch or the spatial relationship between objects. Your mental rehearsal will depend on which senses are most developed in your brain.

For our example of healing back pain, imagine yourself free from pain and doing some sort of exercise or movement you enjoy. See yourself walking agilely, free from pain. Remember, feel the feeling of being pain-free and electrically alive. Imagine the internal and external sensations of your limber back. Feel yourself running free. Choose other sensations that support the healing of your back. If you are sending your intention to heal someone else, carry out all the same aspects of the healing, but imagine yourself inside the other person's back. Send your intention to his back.

Practise Visualizing
You can practise visualization first by getting into a meditative state and imagining the following, while recalling or imagining as much as you can about the sight and smells, and your feelings about them:

- A favourite recent meal (can you remember some of the smells and tastes you really enjoyed?).
- Your bedroom. Walk yourself mentally through it, remembering certain details – the feel of your bedspread, the curtains or carpet. You do not have to see the entire room, just get a detail or impression.
- A recent happy moment (with a loved one, or a child). Remember the most vivid sensations and images.
- Yourself performing an activity such as running, riding a bike, swimming or working out at the gym. Try to feel what it is like for your body to be moving that way.
- Your favourite music (try to 'hear' the music internally).
- A recent experience with an intense physical sensation (such as plunging into a pool or the ocean, having a steam bath,

feeling snow or rain, or making love). Try to relive all of the physical sensations.

To visualize your intention, first work it out carefully ahead of time:

- Now, create a picture in your mind's eye of the desired result. Imagine it as already existing, with you in that situation.
- Try to imagine as much sensory detail as you can about the situation (the look, smell and feel of it).
- Think about it in a positive, optimistic, encouraging way; use mental statements, or affirmations, that confirm that it has or is now happening (not that it will happen in the future). For instance, for someone with a heart problem, 'My heart is healthy and well.'
- For healing, try to imagine healing energy (perhaps as a white light or as your personal deity) filling you and observe it healing the portion of your body that is ill — say, turning a diseased organ into a healthy one. If a good-versus-evil 'contest' is most vivid for you, imagine the 'hero' cells battling or eating up the 'bad guys'. Otherwise, visualize diseased cells or tissue changing into healthy cells, healthy cells replacing diseased cells, or imagine your entire body with that specific body part in perfect health. Visualize yourself often as perfectly healthy, carrying out your daily activities. Find an image of the body part on the Internet or in a book as it looks when it is healthy. Imagine your own body part looking like that.
- If you are in pain, imagine the nerve endings in your entire body and 'see' healing energy being taken in with every breath, flowing through your muscles and blood cells, through your arteries to the nerves, where they are soothed and healed.

- Send out the visualization often, both during meditation and throughout the day.

Belief

The copious evidence of the placebo effect demonstrates the extraordinary power of belief. Belief in the power of intention is also vital. Keep firmly fixed in your mind the desired outcome and do not allow yourself to think of failure. Dismiss any it-won't-happen-to-me type of thoughts. If you are attempting to affect someone who does not share your belief that it may be of benefit, speak to them about some of the scientific evidence in *The Intention Experiment* and elsewhere. It is important that both of you share the same beliefs. Herbert Benson believes that his monks were able to achieve their effects because they used words or phrases incorporating their most deeply held beliefs.[5]

Move Aside

In studies of meditation, mediumship and healing, those who are successful at intention imagine themselves and the person receiving healing as one with the universe. In your meditative state, enter into a zone where you relax your sense of 'I' and sense a merging with the object of your intention and The Field. Frame your intention, state it clearly and then let go of the outcome. At this point, you may sense that the intention is taken over by some greater force. Close your internal meditation with a request and then move your own ego aside. Remember: this 'power' does not originate with you – you are its conduit. Think of it as a request you are sending to the universe.

Timing

The evidence suggests that mind–over–matter intention (that is, psychokinesis) works best at points of increased geomagnetic activity. You can find out about the geomagnetic levels in your area by consulting several websites. The US National Oceanic and Atmospheric Administration (NOAA) created a Space Environment Center (SEC), America's official source of space weather activity (www.sec.noaa.gov). The SEC, in turn, set up a special Space Weather Operations (SWO) branch to act as a warning centre for the world concerning disturbances in space. Jointly operated by the NOAA and the US Air Force, SWO provides forecasts and warnings of solar and geomagnetic activity.

SWO receives its data in real time from a large number of ground-based observatories and satellite sensors around the world. These data enable the SWO to predict solar and geomagnetic activity, and to make worldwide alerts during heavy storms. For the forecast of the day you plan to carry out your intentions, see http://sec.noaa.gov/today2.html.

The SEC has created Space Weather Scales to give lay people an idea of the severity of geomagnetic storms, solar radiation storms and radio blackouts, and their effect on our technological systems (www.sec.noaa.gov/NOAAscales). The numbers attached to them (such as 'G5') indicate the level of severity, with 1 being mild and 5 the most severe.

The Solar and Heliospheric Observatory (SOHO) was set up as a joint project by the European Space Agency and NASA to study the effect of the sun on the earth. For more information, see http://sohowww.nascom.nasa.gov/.

For other aspects of space weather, including charts of geomagnetic activity, see http://sohowww.nascom.nasa.gov/spaceweather/. This website includes useful charts on geomagnetic activity, solar wind and high-energy proton and X-ray flux.

All geomagnetic activity is measured on a K index, with 0 being the most quiet and 9 the most turbulent. The *a* index is similar, but uses a larger scale – from 0 to 400.

When you are sending an intention, plan to do so on a day when the K index is 5 or more (or the *a* index more than 200).

It may also be best to use intention during 1 p.m. local sidereal time (check the web to compute local sidereal time).

Only send intentions on days when you feel happy and well in every way.

Putting It All Together

Your Intention Programme

- Enter your intention space.
- Power up through meditation.
- Move into peak focus through mindful awareness of the present.
- Get onto the same wavelength by focusing on compassion and making a meaningful connection.
- State your intention and make it specific.
- Mentally rehearse every moment of it with all your senses.
- Visualize, in vivid detail, your intention as established fact.
- Time it right – check what the sun is doing, and choose days when you feel happy and well.
- Move aside – surrender to the power of the universe and let go of the outcome.

CHAPTER FOURTEEN

Your Personal Intention Experiments

NOW THAT YOU HAVE PRACTISED 'powering up', what can you use intention for in your own everyday life? To help you find out, with the help of my scientists I have designed a series of informal, personal experiments.

The following 'experiments' are intended to be read in two ways: as a springboard into ways to incorporate intention into your life, and also as a piece of anecdotal research. Whenever you carry out an intention experiment, I would like you to report it on our website.

To carry out these experiments, all you will need in the way of equipment is a notebook and a calendar. When you are first starting, note the date and times of your intentions. Each intention experiment should be carried out after 'powering up' in your intention space, using the programme outlined in chapter 13. Needless to say, if you suffer from a serious illness and are trying to think yourself better, you should augment your own healing intentions with the help of a trained professional healer, whether conventional or alternative.

Make a daily note of any change in the object of your intention, and be specific. If you are trying to heal a condition in yourself or someone else, take a daily 'temperature' of change. What does the person feel like, in general? What symptoms have improved? Have any got worse? Have any new ones turned up? (If any situations seriously worsen, immediately consult a professional practitioner, and also examine any subconscious intentions.)

If you are trying to change your relationship with someone who is ordinarily very antagonistic to something more positive, make a daily note of his or her interactions with you, to determine if anything has changed.

To Have Something Manifest in Your Life

Select a goal that has never happened but that you would like to have happen. Choose something that seldom occurs or is particularly unlikely, so that if it does come to pass it is more likely to be the result of your intention.

Here are some possibilities:

- receiving flowers from your husband (if he has never bought them for you);
- having your wife sit down and watch a football match with you (if she usually refuses to do so);
- having the boorish neighbour who never gives you the time of day start a cheery conversation with you;
- having your child help with the dishes;
- having your child wake up on his or her own in the morning and get ready for school without prompting;
- improving the weather (30 per cent more or less rain, say);
- having your child make his or her bed;
- having your dog stop barking at night;
- stopping your cat from scratching the sofa;

- having your husband or wife come home from work one hour earlier than usual;
- having your child watch television two hours less;
- getting someone who can't stand you at work to say hello and start up a conversation;
- achieving 10 per cent higher profits at work;
- growing your plants or crops 10 per cent faster than usual.

As you begin to manifest, you can try more complicated thoughts. But remember, at first you want one single event to change, something where change can be easily quantified and can probably be attributed to your thoughts.

Retro-intentions
- If you still have a medical problem of some sort, cast your mind back to the point where it started. Carry out an intention for it to resolve itself then. See if you are now better.
- If you are not getting along with someone, cast your mind back to the point where you first had a disagreement and send your intention to change there. Remember to be very specific.
- Ask your friends and family if you can try a retro-prayer for some of their loved ones who were ill 5 years before. Concentrate on their former illness and see if it improves their current state of health. The idea will seem so ridiculous and therefore so harmless that they probably will agree to it. If you feel bold, you may even try this with a local nursing home. First, be sure to obtain the permission of the patient, as well as those in charge.

Report any results by writing in to *The Intention Experiment* website: www.theintentionexperiment.com.

Group Intention Exercises

Assemble a group of your friends who are interested in trying out some group intention exercises. Create an intention space where you will meet each time. Select a group target in your community. Here are a few possibilities:

- improving the weather;
- reducing violent crime by 5 per cent;
- reducing pollution by 5 per cent;
- reducing litter on a particular street in your community;
- getting your mail delivered one hour earlier;
- achieving some form of community activism (such as preventing a mobile phone mast from being built in your area);
- decreasing the incidence of local road accidents involving children by 30 per cent;
- improving the collective grade point average of the local school by one grade;
- decreasing abuse of children in your community by 30 per cent;
- reducing possessions of knives or illegal weapons by 30 per cent;
- increasing (or decreasing) local rainfall by 10 per cent;
- decreasing the number of alcoholics in your area by 25 per cent.

Depending on the nature of your intention, make one member of the group responsible for researching statistics involving your local accident, weather or crime statistics. For these types of statistics, it is a good idea to get hold of reports for the last 5 years in your area and surrounding communities so you have something solid to compare.

Then, when you meet, decide on a group intention statement. When you are 'powering up', visualize yourselves as a single entity (say, a giant bubble or any other unified internal image). Once you are all in a collective meditative state, have one member of the group read out the statement. Meet regularly to send the same intentions. Keep a careful reading of statistics for one month before and several months after you have sent the intentions. Note any changes.

Send the results to *The Intention Experiment* website: www.theintentionexperiment.com.

CHAPTER FIFTEEN

The Group Intention Experiments

YOU ARE NOW INVITED TO PARTICIPATE in massive group intention experiments with many, if not most, of the other readers of this book. If you would like to take part in the largest mind-over-matter experiment in history, read on.

In these group intentions, you will become involved in important new research to further the world's knowledge about the power of intention. There will be blogs and interactive elements on our website, so that you can correspond with like-minded individuals around the world about our results and the results of individual experiments (chapter 14).

Naturally, it is not compulsory. In fact, I would prefer you not to get involved unless you are passionate about participating. I need committed participants, willing to take the intention experiment seriously. Each experiment might take a few minutes to an hour of your time, although in future we might try experiments that take a little longer.

First, log on to the website (www.theintentionexperiment.com). There you will find information about the dates and objectives of future intention experiments. We will plan those dates to coincide with times of a fair degree of geomagnetic activity. Mark those dates in your diary now; if you intend to participate, it is vital that you don't forget. We have a number of experiments planned, but as scientific experiments are expensive to carry out and require lengthy analysis, there will be sizable intervals between experiments. If you miss an intention experiment, you will have to wait a few months for another one.

Several days before the experiment, read through the preliminary instructions to familiarize yourself with what to do. The instructions will explain that you need to carry out many of the 'powering up' exercises of chapter 13 just before you send your intention. You will find information about the time of the experiment in *your* own time zone. The website has a running clock (set to US Eastern Standard Time and Greenwich Mean Time) and a countdown to each new experiment, and will specify the equivalent times in different time zones. Readers around the world will be participating, so it is vital that all the readers send intentions at the right time.

As this is a scientific experiment, we need to have committed and knowledgeable participants, who have read and understood the ideas in this book. Consequently, we will try to weed out potential spoilers or the uncommitted by asking every potential participant to supply a password, which will be taken from phrases or ideas in the book and will vary every few months. We will ask you to supply, for example, the fourth word of the third paragraph on page 57 of the US hardback edition (or page 65 of the paperback). We will make sure we specify passwords for every edition published in every country, so your password will work no matter which version of the book you have read. Just follow the instructions. The only way to be part of the experiment is to have read the book and to log on with the correct password, after which you will be supplied with a private password, to use for future experiments.

Because this is a scientific experiment, we need to know some details about our participants, such as their average age, their gender, their health – or possibly their degree of psychic ability. On the day of the experiment, you will be asked to supply some information about yourself. Several of our scientists have designed short questionnaires for you to fill in. Of course, this information will be kept confidential, under international and national laws of data protection. Once you have filled in our questionnaires, you won't have to re-key any information you have already supplied for any future experiments.

On the day of the intention experiment, at the particular time specified on the website, you will be asked to send a carefully worded, detailed intention, depending on the target site. The website will walk you through the steps. You will be asked to 'power up' into your meditative state, to enter a state of compassion and to send a carefully worded, detailed intention that will be specified on the website.

For instance, let's say that we are trying to send an intention to have a spider plant grow faster at Fritz-Albert Popp's lab in Neuss, Germany, on Friday 20 March at 8 p.m. GMT. We will have a photograph or web camera image of the spider plant on the website, so you can train your intention on the right subject. The website will instruct you to think or say the following sentence on 20 March at 8 p.m.:

Our intention is to have our spider plant in Neuss grow 10 per cent faster than a control plant.

Or, let's say that we have a patient with a wound. Our intention might be:

Our intention is for Lisa's wound to heal 10 per cent faster than normal.

Because this is a scientific experiment, we will structure our experiment to test a precise, carefully quantified result: 10 per cent faster or slower, say, or 6°C cooler than normal or than a control.

Once finished, the results will be analysed by our scientific team – ideally by a neutral statistician as well – and then published on the website.

I must reiterate that I cannot guarantee that the experiments will work – at first or ever. As scientists and objective researchers, we will be duty-bound to faithfully report the data we have. Whether or not our first experiments are successful, we will continue to refine the design with each new experiment as we learn more about group intention. If the first or second or fifth experiment doesn't work, we will keep trying and keep learning more with every result. The nature of frontier science requires that you stumble along blindly, feeling your way along the right path.

Do consult the website frequently for announcements of experiments, postings of the individual experiments (chapter 14) and announcements of the date of every future experiment. If you have enjoyed the written portion of this book, the website will continue the experience for you as an open-ended sequel.

Thoughts Heard 'Round the World: The First Intention Experiments

Since the time of the first release of The Intention Experiment *in hardback, we have run a number of large-scale experiments with a variety of targets online and in front of large audiences. Here are our first results, as of August 2007. For periodic updates, please consult our website: www.theintentionexperiment.com.*

FOR MONTHS I MULLED OVER a possible target for our first large-scale intention experiment. I wanted our target to be philanthropic – something in need of improvement or rescue. The experiment would have to be inspirational, to encourage thousands of readers around the world to participate. The target had to be such that change of any kind could be readily measured. And finally, and possibly most important of all, this first experiment had to be cheap. Scientific experiments can cost thousands of dollars, and I had a limited purse to donate toward the creation of both a website and the experiments.

I put together a scientific team from among some of the leading investigators into consciousness research who'd expressed initial enthusiasm: Dr. Gary Schwartz at the Center for Advances in Consciousness and Health at the University of Arizona, Fritz-Albert Popp in Germany, Robert Jahn and Brenda Dunne at Princeton's PEAR, Dean Radin and Marilyn Schlitz at the Institute for Noetic Sciences in California, physicist Konstantin Korotkov in St. Petersburg, Russia, and British biologist Rupert Sheldrake. The plan was to assemble a consortium of these scientists to discuss protocol together and for different scientists to take turns running the experiments so that I would not be reliant upon the time, resources and results of any single lab.

We began kicking around ideas, but none of us could come up with a workable first experiment satisfying all these initial requirements. Brenda Dunne from the PEAR lab first came up with the idea of the 'ecosphere', filled with tiny plants and primitive animals. We could artificially raise the temperature and then try to lower it through our thoughts. If we were successful, the implications would be enormous: collective thoughts might be able to help cure global warming. But when we began investigating a possible study design, we discovered that an ecosphere would have to be entirely enclosed in order to control moisture and temperature. Unless we had one specially built, the scientists would have an unsolvable problem: the equipment they had available to take measurements of any changes would introduce outside influences (such as air temperature) into the hermetic environment and skew our results.

Gary Schwartz suggested that we attempt to lower the crime rate in Tucson or the mortality rate at a particular hospital. Marilyn Schlitz liked the idea of helping a child with attention deficit become more attentive as attention is a quality that can be easily measured. Dean Radin produced some wonderfully creative ideas that would have turned our website into a computer game: we could create a computer image of someone with Alzheimer's and ask the readers to help him back to his room, observing his progress in real time. Or we could ask readers from different hemispheres of the world to work together to push a computer image in a certain direction – say, toward an icon representing world peace.

Much as I loved every one of these ideas, each presented us with problems of expense or design. I also didn't want to use a human being as one of our first targets; we needed to learn a great deal more about the power of mass intention before we could guarantee a positive effect.

Fortuitously, Gary Schwartz generously volunteered to carry out the first experiment. Gary has a great deal of experience carrying out experiments in energy healing,[1] and he also has a full lab at his disposal, independent funding, and some fantastic equipment. He is the first scientist, after all, to have photographed light emissions from living things through his super-cooled digital CCD camera system, which not only creates digital photos of bio-photon emissions, but analyzes and counts them, one by one. Most recently, he had been studying the effect of distant healing on the light emissions from plant leaves. Like Fritz-Albert Popp, Gary thought we should start by measuring the effect of intention on this current of light, because it is infinitely more subtle than, say, cellular growth rate. The tiniest change in the organism can be controlled for and measured. Besides, Gary said, he likes working with plants, which are nice, clean experimental subjects that don't require the kind of complicated review processes at his university that human beings do.

Gary suggested that we start with a geranium leaf taken from the flourishing plant in the office of his colleague, Dr. Melinda Connor and attempt to replicate the pilot study I'd carried out with Popp. This time, though, we would create a study with a target and an identical control. Both would be subjected to the same conditions, but only one would be sent intention. Although the participants would know our target, the scientists would not be told until they'd calculated the results.

Once more I had to suppress my own disappointment with the suggested target. To persuade thousands to participate in an intention experiment, I told Gary, I needed to engage their hearts and minds; a leaf was hardly going to set the global imagination on fire.

"Remember when the Jodie Foster character in the movie 'Contact' is impatient to fly off into a wormhole?" Gary said. "The other scientist says to her, '*Baby steps, Ellie. Baby steps.*'"

We were about to make scientific history with an experimental design that had never been tried before. We had to start from the ground floor. First we had to establish that the thoughts of a batch of disparate people from around the globe could have an effect – any effect. Only after we'd achieved that could we move on to more ambitious targets.

I had to smile at the idea that we were going to start with a leaf. When the cover design of the US version of this book featuring a leaf was first presented to me, I thought it very beautiful, but a little inappropriate. *What on earth did a leaf have to do with anything?* Now it appeared that a leaf would be our first experimental subject; in fact, it would become the *leitmotif* of *The Intention Experiment. Today a leaf, but tomorrow the world.*

As I was soon to learn, such a simple experiment nevertheless required a 50-step protocol to be painstakingly followed by Gary's lab technician, Mark Boccuzzi. Mark would select two geranium leaves identical in terms of size and number of light emissions. In order to achieve statistical significance, we would need more than thirty data points with which to compare the two leaves. Gary came up with the idea of puncturing each leaf sixteen times. That way, we'd have more than enough data points to compare.

Before we went live on the internet for our first experiment, planned for March 24, 2007, we decided to carry out a trial run using the attendees of an Intention Experiment conference held by my publishing company in London on March 11. Mark would hook up a webcam, and a live image of each leaf would appear on its own web page, visible only to me and my London audience. Just before the experiment, the audience would select one of the leaves. We'd send intention for 10 minutes to the target leaf, after which Mark would place both leaves under the CCD camera to be photographed.

Originally, we planned to have our participants attempt to lower the light emissions, as we had in the Popp experiment. But as March 11 approached, I began to think that an instruction to 'lower light' might confuse people, particularly once I rolled out this experiment on the internet. Most people instinctively would try to increase the light. I phoned Gary a half hour before the experiment was to run at 5 pm UK time, and suggested that we reverse our instruction: the audience should attempt to *increase* the light emissions. Gary agreed and told me to instruct the delegates to use their thoughts to make the leaf 'glow and glow'.

We asked a member of the audience to choose the target leaf by flipping a coin, then displayed the chosen leaf on my Powerpoint projector. After engaging the audience in a simple Powering Up exercise (chapter 13), I asked them to send an intention to make the leaf glow, while playing Choko Rei, a special Reiki chant by Jonathan Goldman that we now use for all our Intention Experiments.

Mark only asked me to reveal which leaf we'd chosen after he'd finished his calculations. Gary phoned me up excitedly a week later. "You won't believe it. The leaf sent intention was glowing so much compared with the other leaf that it seems like the other leaf had a 'neglect effect.'"

In fact, the results of the glowing intention had been so strong that they could readily be seen in the digital images created by the CCD cameras. Numerically, the increased biophoton effect was highly statistically significant. In fact, he said, all the punctured holes in the chosen leaf were filled with light. All the holes in the control leaf, on the other hand, remained black.

Gary wrote up our results for scientific publication, which prevents me at this writing from publishing photos. Science journals demand that all details of a scientific experiment be published first in a peer-reviewed journal before being circulated publicly.

Flush with this first overwhelming success, we geared up for the large experiment on March 24 at 5pm UK time. Gary and his team had hypothesized that an audience required a live image to connect with the target. However, our webmaster cautioned us against using a web cam image, which might jam if thousands of online visitors participated. The next best possibility was a digital camera that could take refreshed images every few seconds. Mark had such two such cameras, which could be set up to feed our website with a continually refreshed image of both leaves every 15 seconds.

A half hour before the experiment was to begin, I called in my 10-year-old daughter, Anya. As this book is dedicated to this master of intention, it was appropriate that she be the one to flip a coin and choose the leaf. Then we sat in front of our iMac at home, like the many thousands we expected to participate.

When 5 pm arrived, we couldn't access the site. We tried again and again for the next forty minutes – well beyond the time the experiment should have finished. If we couldn't get in, it was highly likely that no one else could, either. We called our webmaster and he confirmed what we already knew: the website had crashed. We were victims of our success. An estimated 10,000 people attempted to participate in the experiment, he said, which overwhelmed the site.

Many people wrote in to say they'd tried to send intention anyway. Indeed, after Gary and his team finished their analysis, the figures showed a bit of movement in the predicted direction. But with so many people confused and unable to access the experiment, the results were inconclusive.

It began to dawn on me that the biggest challenge with these experiences was not in demonstrating the power of intention but in finding an internet system sophisticated enough to allow thousands of people around the world to stare at the same 'live' image at the same time. I'd spent so much time choosing the experimental targets that I'd never considered the technical issues involved in having thousands of people opening the same web page at exactly the same moment. As I was to discover over the next few days, allowing in such sizeable simultaneous traffic required a vast amount of extra web capacity.

We have a countdown clock on our website, to announce the number of days, hours and minutes before our next experiment. With so many disappointed readers, I reset the countdown for April 14, committing myself to just three weeks to come up with a solution to our special problems.

We hired a team of web designers called Visionwt. Tony Wood, the managing director, convinced us that to ease the huge surge of web traffic created during the experiments, it was best to hold the experiments on a special page, away from the main website. Tony also proposed to control the flipping over of pages, rather than having readers click to other pages themselves, so there would be no possibility of the site freezing when everyone clicked the same button at the same time.

The ability of a website to handle simultaneous web traffic is completely reliant upon the size of a web system's server power. To make this work, we were going to need an enormous amount of server capacity on hand. On Tony's recommendation, we rented server space from a company that supplies the servers for Pop Idol. We were taking no chances. These were people well versed in preventing a massive cyber traffic jam.

Tony and his team suggested that we run two experiments – a test on April 14 for the real one on April 21. The scientists were still operating from the assumption that it was important to have a 'live' connection with the target and so designed the event to have the continually refreshing photo. This time, for variation, Gary suggested that we use string bean seeds as our target to 'glow.'

All our planning finally paid off. Nearly 7000 people from thirty countries participated in the experiment, and only a handful had problems logging on – with our team on hand to sort them out. Nine linked servers were on hand to distribute the load. For the first few moments of our experiment, they were almost full.

This time, after Gary had analyzed the results, the bean seed experiment showed a strong 'glow effect' – the same as the London leaf experiment – but not in terms of statistical significance, largely because of the limitations of the imaging equipment.

"The beans were in the predicted direction, but the results did not reach statistical significance," Gary wrote to me in an email. "However, there were only 12 beans per condition (glow versus control). If it was possible to image twice as many beans, the results would have reached statistical significance (through what is called power analysis in statistics)."

In other words, we showed a large effect, but we needed more seeds just to satisfy the scientific definition of 'significant.'

For a few moments our readers believed that they'd actually seen the beans glowing. As part of the protocol, once we begin sending intention, I would call Mark – the cue for him to turn on a fluorescent light. I was so preoccupied making sure that the technology worked properly that I forgot to tell Mark to turn on the light. Two minutes into the experiment, I suddenly remembered and phoned him. As this was a live image, no sooner had Mark received the instruction than the beans appeared to be lit with light on our web page. Many readers wrote in our forum later that they'd seen the seeds actually glowing. I had to break the news that this particular glow effect had a more prosaic cause.

A week later, on April 21 we ran what we considered the first real experiment – this time with our geranium leaf and again, with 'glow' instructions. Yet, although some 7000 to logged on to the main site, only 500 managed to get into the experimental portal.

This time, we had little effect. Less than a twelfth of the number of people who participated in the leaf experiment had been able to participate in the leaf experiment, which may have been one reason why the results were inconclusive. There was also the possibility that mass intention required a threshold – a critical mass of participants – in order to have an effect.

These first pilot experiments left us with a few tantalizing preliminary hypotheses. Intention sent by a group of just a few hundred occupying the same room had had a significant effect on a distant target – as large an effect as a group of nearly 7000 people sending intention from remote sites around the world. The failed April 21 study suggested the possibility that we needed a certain size 'dose' of intention to have an effect. The Transcendental Meditation organization has long held a similar view of meditation – that a certain percentage of regular meditators is required to exert an effect on misery indices such as the crime rate.

It was also evident that our technical problems had interfered with intention. In the March 24 and April 21 experiments, the technical glitches may have caused the experiments to fail. During the March 24 experiment, although many people had tried to send intention to a mental image of the geranium leaf, we did not record a significant effect. Gary and I also wondered whether group intention had an 'enhanced' effect when carried out by a focused, coherent group in the same physical space. But at this point, these were only tantalizing possibilities that would have to be tested out over many experiments before we could hold to them with any certainty. The lack of an effect on the April 21 study could have been due either to the small numbers participating or to technical problems. At this point, it was just too early to call.

I was getting a little discouraged at the technological difficulties. The website constructed for these experiments was just not up to the job of mass intention and many of the solutions suggested by other webmasters were impossibly expensive. The technology of the second experiment had worked, but afterward we'd been presented with an extraordinarily large bill. The server power alone had cost us $6,000 for a half hour and the special web pages many thousands more – far too much for me to donate on a regular basis.

While we were still struggling to find our way through all these technical challenges, something remarkable occurred on our site. A

Vietnam vet named Don Berry wrote in to the website forum, offering to be our first human intention experiment. His spine was fused,
making it impossible for him to move from side to side. As he had a
wealth of x-rays and other medical test reports, he could produce a
full record of his medical history by which to measure any change.

Don's blog prompted our other readers to set twice-weekly periods
during which they would send intention to Don, and he in turn
began to keep a diary of his condition. After a few weeks, he wrote in
to say that he was feeling better. His progress wasn't scientific in any
sense, just a personal intention experiment, but it offered a wonderful example of the kind of personal intention experiment that readers
could try in their own lives.

Don's experience also sparked an idea. Perhaps we could run regular, informal experiments for people like Don – an Intention of the
Week. I put up a photo of a man with multiple sclerosis and invited
my readers to send intention to him on the following Saturday and
also to nominate someone else in need. Before long I was receiving
dozens of requests: people with cancer, children with brain damage
or birth defects, estranged family members, wounded pets. I tried to
choose carefully and always put up a photo of our Intention of the
Week: a child with cerebral palsy; a newborn who'd sustained brain
damage after a difficult birth; a teenage runaway; a woman who'd
found the love of her life in late middle age only to discover she'd got
breast cancer. I asked the remote viewers in the audience to find
Madeleine McCann, the British four year old who'd been abducted
in Portugal, and got back extraordinarily detailed responses, which I
forwarded to the police. Our site was turning into the cyber equivalent of a weekly prayer group.

The most remarkable success at this writing is helping to reunite a
teenaged runaway with her mother. A sixteen-year-old named
Briana had left home. Her mother feared that she was under the
influence of a rough crowd of lesbian students since she was
pretending to be gay. The mother was not anti-gay, but felt her

daughter was unduly influenced by this group and not being true to herself. We sent an intention for Briana and her mother to communicate more honestly. After several weeks I received an effusive note from her mother. Briana had come home three weeks after the intentions started and they'd begun having honest, heartfelt talks. She admitted that she was not really gay and disclosed that her problems in trusting men stemmed from her relationship with her father.

Not all targets were successful; Mel Taylor, the newlywed with breast cancer, died three days before we carried out our intention; a young woman with cerebral palsy recorded no change.

Nevertheless, the site soon became a beacon for healing; when Drew, a baby who'd sustained brain damage, was chosen as the Intention of the Week, four healers, including a Hawaiian Kahuna healer, wrote in to his parents to offer special healing during our intention. Just the thought of becoming an Intention of the Week, even before he'd been posted on our site, seemed to have helped Vidar Nilson sail through surgery with better than anticipated results.

In our scientific intention experiments, we had established very preliminary evidence that group intention could have an effect on a minuscule biological process. By June 2007, it was time to move up a gear – to send intention to something more representative of a real-life setting. Gary, Mark and I began kicking around the idea of testing whether intention could affect the growth of plants. There'd been some interesting data on seeds and intention; Canadian psychologist Bernard Grad had carried out several studies showing that seeds irrigated with water held by a healer had a faster germination rate and growth than controls.[2] Then, British researcher Serena Roney-Dougal and parapsychologist Jerry Solfvin tested whether healing intention could be used to affect the health and growth of plants on a commercial organic farm. They'd chosen lettuce plants because they germinate rapidly and so would allow for a number of trials. This time, however, the healer was asked to send intention directly to the seeds themselves. In their first study, the seeds given intention didn't sprout any faster or grow larger, but they were healthier and had less fungal disease and slug damage than the controls.[3] In a replication study, however, they'd showed enhanced growth, as well as health.[4]

Although we wanted to base the design of our experiment on Roney-Dougal's study, Mark recommended that we use barley seeds – the food of choice of most livestock, and a healthy grain for humans. Like the lettuce study, Gary and Mark planned to prepare four sets of seeds – one set of seeds, and three controls – to eliminate chance findings. As the continually refreshing photo had been a major factor in our need for so much server power in the earlier studies, I asked if we could use a simple photograph of the target with these new studies. An ordinary photo had seemed to work with our Fritz Popp pilot study in London; indeed, with the algae portion of the study, I'd simply shown the study participants a picture of *some* algae I'd photographed when visiting Popp's office, not *the* algae we'd actually sent intention to. Perhaps a photo – indeed, any photo – enabled the mind to connect with the target.

I was scheduled to appear before many diverse audiences in many countries during the summer of 2007, which gave me numerous opportunities to test this new experiment in a variety of settings. We scheduled our trial run for an audience of 500 in Australia. Mark sent me all four photos of seeds, each nestled in a seed pocket; right before our experiment was to begin I asked a member of the audience to select our target randomly. We'd tried to model the wording of intention closely to that used by Roney-Dougal in her experiments – for the target seeds 'to enjoy enhanced germination, greater growth and greater health'. After our 10-minute intention, Mark planted the seeds and during the next two weeks took regular measurements.

Before we even had our first results, I wanted to run the experiment on our website. But first we had to figure out how to conduct this experiment on the web at an affordable price. For weeks I'd advertised our next Intention Experiment for July 7. It was late June and I didn't want to let the readers down, but with only two weeks to go I began to despair of ever clearing what seemed to be an impossible technological hurdle. Finally, I decided to take my own advice. I cleared my mind and set an intention – to find the perfect webmaster and solution.

A few days later, I received a phone call from Guus Goris, a Dutch friend of mine, who informed me that a group of computer executives might be able to help us without charge. He also introduced me to Nick Haenen, a web designer with his own spiritual website. Nick had come up with an ingenious solution to our need for vast server power. Instead of renting our own servers, Nick said, why not make use of the giant capacity already created by a social network portal, like MySpace or Facebook? He'd constructed some sites using the Ning social network portal. Ning offers individual organizations instant facilities for a community-based website. The main advantage of Ning, for our purposes, was its server capacity – some 500 linked servers – to cope with the organization's 20,000 social networks.

When Nick contacted the Ning creators they were enthusiastic about using their equipment to run intention experiments (Ning, by the way means 'love' in Chinese), and began working with Nick to modify the system slightly to cope with our special needs. By the time they were through, they said, our system would be able to cope with a hundred thousand simultaneous users.

The new site had only been up for five days when we ran the germination experiment on the web, at another temporary web address. As my regular audience had no advance warning of the new site, only 500 people participated. No matter what the scientific results would yield, the experiment had been an overwhelming success, as far as I was concerned: the site had stayed up, even though our instructions required that all the participants click onto the same page at the same exact moment.

After a few weeks, Mark wrote to me to say that in both instances our intention did not produce a measurable effect. His preliminary analysis of the Australian study and our internet study seemed to show no difference between the target seeds and the controls. Gary worried that the two studies suffered from intention 'contamination'. To ensure that all four sets of seeds were planted in conditions with the same light, moisture and soil conditions, Mark had planted the target and the controls in the same dirt. As the seeds had not been shielded from each other, there also could have been 'biophoton' contamination. In other words, the seeds sent intention could have *communicated* that intention to the control seeds via light emissions.

The greatest challenge of the scientific method is determining why something works or why it fails. A failure can suggest myriad possibilities. The failure of these experiments could mean that intention cannot make healthy seeds germinate or grow any quicker than normal, or that we can't make a 'healthy' system healthier (we found that in a number of studies recounted in *The Intention Experiment*). Or perhaps we had not made our intention specific enough. All we had specified was that the plants grow 'faster and be healthier than normal'.

Or it could be, as Gary suspected, that the physical design of our experiment contaminated the results through light emissions or even shared space. Far fewer people had participated than normal, so perhaps we didn't have a large enough critical mass of people involved in the experiment to register an effect. When venturing out into this kind of virgin territory, the most outlandish possibility must nevertheless be entertained.

We decided to try the experiment again at a workshop I was holding at in Rhinebeck, New York – this time with a more specific intention, and with all the seed samples isolated from each other, so they would not be able to 'share information'. Melinda Connor recommended that we change our intention to a highly specific one: 'My intention is for all the seeds in our target group to sprout at least *three* inches by the *fourth* day of growing.'

A few weeks later, when Gary was analyzing the data from all three studies on a graph, he noticed some strange spikes. On closer analysis, he discovered that some 10 per cent of the seeds in each group didn't sprout. From this he realized that the ordinary statistical method he'd been using didn't apply to these figures. If the distribution of the figures is not normal, but deviates from a bell-shaped curve, the more accurate means of analyzing them is through non-parametric statistics, which don't require a normal distribution. Gary re-analyzed the data, using two types of non-parametric statistical methods to examine the combination of Experiments 1 and 2, and two methods for analyzing Experiment 3 on its own.

On August 2, he wrote me excitedly to say that he'd come up with some amazing results. For each of the three experiments, the germinated intention seedlings were longer than the control seedlings. A graph he'd produced of the data showed this in bold relief – analyzed together, all the intention seeds consistently grew larger than the controls.

The third study had to be analyzed on its own, as the design was different – with a longer growth time (six hours), a more focused intention (a specific growth instruction) and separate seed group growing conditions. According to his calculations, the Rhinebeck study had generated *significantly larger seed growth overall* (in both intention and control seeds) than the other two.

The combined results of experiments 1 and 2 showed a significant effect with one set of the statistics, while experiment 3 showed a significant effect with both sets of statistics.

"I suspect that Run 3 is superior to Runs 1 and 2 for a host of reasons, including more focused intention instruction, longer growing time and separation of seeds by conditions so that potential energetic and intention contamination or cross communication in the water and via biophotons are minimized," Gary wrote to me.

This presented us with even more intriguing possibilities. A group of 500 people scattered around the globe produced the same effect as a group of the same size sitting together but located halfway around the world. According to this data, there was no threshold; even a tiny group of 100 people in a room in upstate New York had been able to profoundly affect a batch of seeds 3000 miles away.

Now that the technological problems appear to be solved, we will be moving on from beans and leaves. We are at work planning our first simple human study, as well as our ecosphere, and a first experiment to attempt to change the pH of water, which will have vast implications for our ability to clean up polluted lakes or seas. I'm also in discussion with Deepak Chopra and his Alliance for New Humanity to create an intention experiment to lower conflict in one of the hotspots in the world. Several companies have come forward to request that we run an intention experiment within their organizations to lower absenteeism. We'll also examine further whether a critical mass of people is needed in group intention or whether it works best when participants occupy the same physical space.

When you conduct a scientific study, you roam across new terrain a little aimlessly without a compass. Once you find your destination, it usually isn't the one you were looking for. Flexibility is the greatest prerequisite of a good scientist.

Each new intention experiment offers one tiny piece of this unfolding, new scientific story. Although these first experiments have 'worked,' it's important to understand that all we have at the moment is an intriguing demonstration of possibility, and not one single definitive statement. Each scientific experiment must be replicated many times to be accepted as fact, which is why we regularly repeat every experiment in many groups, large and small. Each time, we take one more baby step forward. With every answer – no matter what that is – we will keep learning, and so will you.

www.theintentionexperiment.com

Notes

Preface

1. N. Hill, *Think and Grow Rich: The Andrew Carnegie Formula for Money Making*, New York: Ballantine Books (reissue edn), 1987.
2. J. Fonda, *My Life So Far*, London: Ebury Press, 2005: 571.

Introduction

1. For a complete description of these scientists and their findings, consult L. McTaggart, *The Field: the Quest for the Secret Force of the Universe*, London: HarperCollins, 2001.
2. The full title of Newton's major treatise is *Philosophiae Naturalis Principia Mathematica*, a name that offers a nod to its philosophical implications, although it is always referred to reverentially as the *Principia*.
3. R. P. Feynman, *Six Easy Pieces: The Fundamentals of Physics Explained*, London: Penguin, 1995: 24.
4. McTaggart, *The Field*, op. cit.
5. Eugene Wigner, the Hungarian-born American physicist who received a Nobel Prize for his contribution to the theory of quantum physics, is one of the early pioneers of the central role of consciousness in determining reality and argued, through a thought experiment called 'Wigner's friend', that the observer, 'the friend', might collapse Schrödinger's famous cat into a single state or, like the cat itself, remain in a state of superposition until another 'friend' comes into the lab. Other proponents of 'the observer effect' include John Eccles and Evan Harris Walker. John Wheeler is credited with espousing the theory that the universe is participatory: it only exists because we happen to be looking at it.
6. McTaggart, *The Field*, op. cit.
7. E. J. Squires, 'Many views of one world – an interpretation of quantum theory', *European Journal of Physics*, 1987; 8: 173.
8. B. F. Malle et al., *Intentions and Intentionality: Foundations of Social Cognition*, Cambridge, Mass.: MIT Press, 2001.
9. M. Schlitz, 'Intentionality in healing: mapping the integration of body, mind, and spirit', *Alternative Therapies in Health and Medicine*, 1995; 1 (5): 119–20.
10. R. G. Jahn et al., 'Correlations of random binary sequences with prestated operator intention: a review of a 12-year program', *Journal of Scientific Exploration*, 1997; 11: 345–67.
11. R. G. Jahn et al., 'Correlations of random binary sequences', op. cit.; Dean Radin and Roger Nelson, 'Evidence for consciousness-related anomalies in random physical systems', *Foundations of Physics*, 1989; 19 (12): 1499–514; McTaggart, *The Field*, op. cit.: 116–17.
12. These studies are itemized in great detail in D. Benor, *Spiritual Healing*, Volume 1, Southfield, Mich.: Vision Publications, 1992.
13. Rene Peoc'h, 'Psychokinetic action of young chicks on the path of a "illuminated source"', *Journal of Scientific Exploration*, 1995; 9 (2): 223; R. Peoc'h, 'Chicken

imprinting and the tychoscope: An Anpsi experiment', *Journal of the Society for Psychical Research*, 1988; 55: 1; R. Peoc'h, 'Psychokinesis experiments with human and animal subjects upon a robot moving at random', *The Journal of Parapsychology*, September 1, 2002.

14. William G. Braud and Marilyn J. Schlitz, 'Consciousness interactions with remote biological systems: anomalous intentionality effects', *Subtle Energies and Energy Medicine*, 1991; 2 (1): 1–27; McTaggart, *The Field*, op. cit.: 128–9.

15. Marilyn Schlitz and William Braud, 'Distant intentionality and healing: assessing the evidence', *Alternative Therapies in Health and Medicine*, 1997; 3 (6): 62–73.

16. William Braud and Marilyn Schlitz, 'A methodology for the objective study of transpersonal imagery', *Journal of Scientific Exploration*, 1989; 3 (1): 43–63.

17. W. Braud et al., 'Further studies of autonomic detection of remote staring: replication, new control procedures and personality correlates', *Journal of Parapsychology*, 1993; 57: 391–409; M. Schlitz and S. LaBerge, 'Autonomic detection of remote observation; two conceptual replications', in D. Bierman (ed.), *Proceedings of Presented Papers: 37 Annual Parapsychological Association Convention*, Amsterdam, Fairhaven, Mass.: Parapsychological Association, 1994: 465–78.

18. D. Benor, *Spiritual Healing: Scientific Validation of a Healing Revolution*, Southfield, Mich.: Vision Publications, 2001.

19. F. Sicher, E. Targ et al., 'A randomized double-blind study of the effect of distant healing in a population with advanced AIDS: report of a small scale study', *Western Journal of Medicine*, 1998; 168 (6): 356–63. For a full description of the studies, see McTaggart, *The Field*, op. cit.: 181–96.

20. Psychologist Dean Radin conducted a meta-analysis in 1989 at Princeton University of all known dice experiments (73) published between 1930 and 1989. They are recounted in his book *Entangled Minds*, New York: Paraview, 2006: 148–51.

21. J. Hasted, *The Metal Benders*, London: Routledge & Kegan Paul, 1981, as cited in W. Tiller, *Science and Human Transformation; Subtle Energies, Intentionality and Consciousness*, Walnut Creek, Calif.: Pavior Publications, 1997: 13.

22. McTaggart, *The Field*, op. cit.: 199.

23. W. W. Monafo and M. A. West, 'Current recommendations for topical burn therapy', *Drugs*, 1990; 40: 364–73.

Chapter 1: Mutable Matter

1. All personal information about Tom Rosenbaum and Sai Ghosh and their studies have been culled from multiple interviews conducted in February and March 2005.

2. This was the solution posed by Giorgio Parisi at Rome in 1979.

3. S. Ghosh et al., 'Coherent spin oscillations in a disordered magnet', *Science*, 2002; 296: 2195–8.

4. Once again, I am indebted to Danah Zohar for her easy-to-digest description of quantum non-locality, which appears in D. Zohar, *The Quantum Self*, London: Bloomsbury, 1991: 19–20.

5. A. Einstein, B. Podolsky and N. Rosen, 'Can quantum-mechanical description of physical reality be considered complete?' *Physical Review*, 1935; 47: 777–80.

6. A. Aspect et al., 'Experimental tests of Bell's inequalities using time-varying analyzers', *Physical Review Letters*, 1982; 49: 1804–7; A. Aspect, 'Bell's inequality test: more ideal than ever', *Nature*, 1999; 398: 189–90.

7. Science Fact: Scientists achieve 'Star Trek'-like feat – The Associated Press, December 10, 1997, posted on CNN, http://edition.cnn.com/TECH/9712/10/beam.me.up.ap.

8. Non-locality was considered to be proven by Aspect et al.'s experiments in Paris in 1982.

9. J. S. Bell, 'On the Einstein-Poldolsky-Rosen paradox', *Physics*, 1964; 1: 195–200.

10. S. Ghosh et al., 'Entangled quantum state of magnetic dipoles', *Nature*, 2003; 435: 48–51.

11. Details of Vedral's views and experiments the result of multiple interviews, February, October and December 2005.

12. C. Arnesen et al., 'Thermal and magnetic entanglement in the 1D Heisenberg Model', *Physical Review Letters*, 2001; 87: 017901.

13. V. Vedral, 'Entanglement hits the big time', *Nature*, 2003; 425: 28–9.

14. T. Durt, interview with author, April 26, 2005.

15. B. Reznik, 'Entanglement from the vacuum', *Foundations of Physics*, 2003; 33: 167–76; Michael Brooks, 'Entanglement: The weirdest link', *New Scientist*, 2004; 181 (2440): 32.

16. John D. Barrow, *The Book of Nothing*, London: Jonathan Cape, 2000: 216.

17. Erwin Laszlo, *The Interconnected Universe: Conceptual Foundations of Transdiscipinary Unified Theory*, Singapore: World Scientific Publishing, 1995: 28.

18. A. C. Clarke, 'When will the real space age begin?' *Ad Astra*, May–June 1996; 13–15.

19. Harold Puthoff, 'Ground state of hydrogen as a zero–point-fluctuation-determined state', *Physical Review D*, 1987; 35: 3266.

20. B. Haisch, Alfonso Rueda and H. E. Puthoff, 'Inertia as a zero–point-field Lorentz force', *Physical Review A*, 1994; 49 (2): 678–94; Bernhard Haisch, Alfonso Rueda and H. E. Puthoff, 'Physics of the zero–point field: implications for inertia, gravitation and mass', *Speculations in Science and Technology*, 1997; 20: 99–114.

21. Various interviews with Hal Puthoff, 1999–2000.

22. Reznik, 'Entanglement from the vacuum', op. cit.

23. McTaggart, *The Field*, op. cit.: 35–6.

24. J. Resch et al., 'Distributing entanglement and single photons through an intra-city, free-space quantum channel', *Optics Express*, 2005; 13 (1): 202–9; R. Ursin et al., 'Quantum teleportation across the Danube', *Nature*, 2004; 430: 849.

25. M. Arndt et al., 'Wave–particle duality of C60 molecules', *Nature*, 1999; 401: 680–2; doi: 10.1038/44348.

26. A. Zeilinger, 'Probing the limits of the quantum world', *Physics World*, March 2005 (online journal: http://www.physicsweb.org/articles/world/18/3/5/1).

Chapter 2: The Human Antenna

1. All personal details about Gary Schwartz and his discoveries result from multiple interviews with him and the author, March–June 2006.

2. H. Benson et al., 'Decreased systolic blood pressure through operant conditioning techniques in patients with essential hypertension', *Science*, 1971; 173 (3998): 740–2.

3. E. E. Green, 'Copper wall research psychology and psychophysics: subtle energies and energy medicine: emerging theory and practice', *Proceedings, First Annual*

Conference, International Society for the Study of Subtle Energies and Energy Medicine (ISSSEEM), Boulder, Colorado, 21–25 June 1991.

4. This research was eventually published as G. Schwartz and L. Russek, 'Subtle energies – electrostatic body motion registration and the human antenna-receiver effect: a new method for investigating interpersonal dynamical energy system interactions', *Subtle Energies and Energy Medicine*, 1996; 7 (2): 149–84.

5. E. E. Green et al., 'Anomalous electrostatic phenomena in exceptional subjects', *Subtle Energies and Energy Medicine*, 1993; 2: 69; W. A. Tiller et al., 'Towards explaining anomalously large body voltage surges on exceptional subjects, Part I: The electrostatic approximation', *Journal of Scientific Exploration*, 1995; 9 (3): 331.

6. William A. Tiller, 'Subtle energies', *Science & Medicine*, 1999, 6 (3): 28–33.

7. A. Seto et al., 'Detection of extraordinary large biomagnetic field strength from the human hand during external qi emission', *Acupuncture and Electrotherapeutics Research International*, 1992; 17: 75–94; J. Zimmerman, 'New technologies detect effects in healing hands', *Brain/Mind Bulletin*, 1985; 10 (2): 20–3.

8. B. Grad, 'Dimensions in "Some biological effects of the laying on of hands" and their implications', in H. A. Otto and J. W. Knight (eds.), *Dimension in Wholistic Healing: New Frontiers in the Treatment of the Whole Person*, Chicago: Nelson-Hall, 1979: 199–212.

9. L. N. Pyatnitsky and V. A. Fonkin, 'Human consciousness influence on water structure', *Journal of Scientific Exploration*, 1995; 9 (1): 89.

10. G. Rein and R. McCraty, 'Structural changes in water and DNA associated with new physiologically measurable states', *Journal of Scientific Exploration*, 1994; 8 (3): 438–9.

11. W. Tiller would eventually write about the effect of shielding psychics in his book *Science and Human Transformation*, Walnut Creek, Calif.: Pavior Publishing, 1997: 32.

12. M. Connor, G. Schwartz et al., 'Oscillation of amplitude as measured by an extra low frequency magnetic field meter as a biophysical measure of intentionality'. Paper presented at the Toward a Science of Consciousness Conference, Tucson, Arizona, April 2006.

13. Sicher, Targ et al., 'A randomized double-blind study', op. cit.

14. See McTaggart, *The Field*, op. cit.: 39, for a full description of F.-A. Popp's earlier work.

15. S. Cohen and F.-A. Popp, 'Biophoton emission of the human body', *Journal of Photochemistry and Photobiology*, 1997; 40: 187–9.

16. K. Creath and G. E. Schwartz, 'What biophoton images of plants can tell us about biofields and healing', *Journal of Scientific Exploration*, 2005; 19 (4): 531–50.

17. S. N. Bose, 'Planck's *Gesetz und Lichtquantenhypothese*', *Zeitschrift für Physik*, 1924; 26: 178–81; A. Einstein, 'Quantentheorie des einatomigen idealen Gases [Quantum theory of ideal monoatomic gases]', *Sitz. Ber. Preuss. Akad. Wiss.* (Berlin), 1925; 23: 3.

18. C. E. Wieman and E. A. Cornell, 'Seventy years later: the creation of a Bose-Einstein condensate in an ultracold gas', *Lorentz Proceedings*, 1999; 52: 3–5.

19. K. Davis et al., 'Bose-Einstein condensation in a gas of sodium atoms', *Physical Review Letters*, 1995; 75: 3969–73.

20. M. W. Zwierlein et al., 'Observation of Bose-Einstein condensation of molecules', *Physical Review Letters*, 2003; 91: 250401.

21. H. Fröhlich, 'Long range coherence and energy storage in biological systems', *Int. J. Quantum Chem.*, 1968; II: 641–9.
22. For this entire example, see Tiller, *Science and Human Transformation*, op. cit.: 196.
23. M. Jibu et al., 'Quantum optical coherence in cytoskeletal microtubules: implications for brain function', *Biosystems*, 1994; 32: 195–209; S. R. Hameroff, 'Cytoplasmic gel states and ordered water: possible roles in biological quantum coherence', *Proceedings of the 2nd Annual Advanced Water Sciences Symposium*, Dallas, Texas, 1996.

Chapter 3: The Two-Way Street

1. For all history of Cleve Backster's discoveries and experiments, interview with Backster, October 2004 and his *Primary Perception: Biocommunication with Plants, Living Foods, and Human Cells*, Anza, Calif.: White Rose Millennium Press, 2003.
2. As Obi-Wan Kenobe tells Luke Skywalker, after Alderan has been blown up by the Empire in *Star Wars* part IV: *A New Hope*: 'I feel a great disturbance in the Force. As if millions of voices suddenly cried out in terror, and were suddenly silenced.'
3. Presentation given at the Tenth Annual Parapsychology Association meeting in New York City, September 7, 1967. Also published as C. Backster, 'Evidence of a primary perception in plant life', *International Journal of Parapsychology*, 1968; 10 (4): 329–48.
4. P. Dubrov and V. N. Pushkin, *Parapsychology and Contemporary Science*, New York and London: Consultants Bureau, 1982.
5. P. Tompkins and C. Bird, *The Secret Life of Plants*, New York: Harper & Row, 1973.
6. 'Boysenberry to Prune, Boysenberry to Prune: Do you read me? Lie detector expert Cleve Backster reported in the annual meeting of the American Association for the Advancement of Science that he had detected electrical impulses between two containers of yogurt at opposite ends of his laboratory. Backster claims the bacteria in the containers were communicating.' *Esquire*, January 1976.
7. Backster, 'Evidence of a primary perception', op. cit.
8. Backster, *Primary Perceptions*, op. cit.: 112–13.
9. Backster, *Primary Perceptions*. See also Rupert Sheldrake, *Dogs That Know When Their Owners Are Coming Home and Other Unexplained Powers of Animals*, London: Three Rivers Press, 2000.
10. This and other personal details of events resulted from interviews with Ingo Swann, New York, July 2005.
11. See McTaggart, *The Field*, op. cit.: 39 for a full description of F.-A. Popp's earlier work.
12. All details of these experiments resulted from an interview between the author and Fritz-Albert Popp, January 2006.
13. R. M. Galle et al., 'Biophoton emission from *Daphnia magna*: A possible factor in the self-regulation of swarming', *Experientia*, 1991; 47: 457–60; R. M. Galle, 'Untersuchungen zum dichte und zeitabhängigen Verhalten der ultraschwachen Photonenemission von pathogenetischen Weibchen des Wasserflohs Daphnia magna.' Dissertation. Universität Saarbrücken, Fachbereich Zoologie, 1993.
14. F.-A. Popp et al., 'Nonsubstantial biocommunication in terms of Dicke's Theory', in M. W. Ho, F.-A. Popp and U. Warnke (eds.), *Bioelectrodynamics and Biocommunication*, Singapore: World Scientific Publishing, 1994: 293–317; J. J.

Chang et al., 'Research on cell communication of *P. elegans* by means of photon emission', *Chinese Science Bulletin*, 1995; 40: 76–9.

15. J. J. Chang et al., 'Communication between Dinoflagellates by means of photon emission', in L. V. Beloussov and F.-A. Popp (eds.), *Proceedings of International Conference on Non-equilibrium and Coherent Systems in Biophysics, Biology and Biotechnology, Sep. 28–Oct. 2 1994*, Moscow: Bioinform Services Co., 1995: 318–30.

16. Interview with Popp, Neuss, Germany, March 1, 2006.

17. F.-A. Popp et al., 'Mechanism of interaction between electromagnetic fields and living organisms', *Science in China* (Series C), 2000; 43 (5): 507–18.

18. Ibid.

19. L. Beloussov and N. N. Louchinskaia, 'Biophoton emission from developing eggs and embryos: Nonlinearity, wholistic properties and indications of energy transfer', in J. J. Chang et al. (eds.), *Biophotons*, London: Kluwer Academic Publishers, 1998: 121–40.

20. K. Creath and G. E. Schwartz, 'What biophoton images of plants can tell us about biofields and healing', *Journal of Scientific Exploration*, 2005; 19 (4): 531–50.

21. A. V. Tschulakow et al., 'A new approach to the memory of water', *Homeopathy*, 2005; 94: 241–7.

22. E. P. A. Van Wijk and R. Van Wijk, 'The development of a bio-sensor for the state of consciousness in a human intentional healing ritual', *Journal of International Society of Life Information Science (ISLIS)*, 2002; 20 (2): 694–702.

23. M. Connor, 'Baseline testing of energy practitioners: Biophoton imaging results.' Paper presented at the North American Research in Integrative Medicine conference, Edmonton, Canada, May 2006.

24. Personal details about K. Korotkov the result of multiple interviews with the author, November–March 2005–2006.

25. S. D. Kirlian and V. K. Kirlian, 'Photography and visual observation by means of high frequency currents', *J. Sci. Appl. Photogr.*, 1964; 6: 397–403.

26. Korotkov's most important work on the subject was K. Korotkov, *Human Energy Field: Study with GDV Bioelectrography*, New Jersey: Backbone Publishing Co., 2002; K. Korotkov, *Aura and Consciousness – New Stage of Scientific Understanding*, St Petersburg: St Petersburg Division of the Russian Ministry of Culture, State Publishing Unit 'Kultura', 1999.

27. K. Korotkov et al., 'Assessing biophysical energy transfer mechanisms in living systems: The basis of life processes', *The Journal of Alternative and Complementary Medicine*, 2004; 10 (1): 49–57.

28. L. W. Konikiewicz and L. C. Griff, *Bioelectrography – A new method for detecting cancer and body physiology*, Harrisburg, Va.: Leonard Associates Press, 1982; G. Rein, 'Corona discharge photography of human breast tumour biopsies', *Acupuncture & Electrotherapeutics Research*, 1985; 10: 305–8; K. Korotkov et al., 'Stress diagnosis and monitoring with new computerized "Crown-TV" device', *Journal of Pathophysiology*, 1998; 5: 227.

29. P. Bundzen et al., 'New technology of the athletes' psycho-physical readiness evaluation based on the gas-discharge visualisation method in comparison with battery of tests', *'SIS-99' Proceedings, International Congress St Petersburg*, 1999: 19–22; P. V. Bundzen, et al., 'Psychophysiological correlates of athletic success in athletes training for the Olympics', *Human Physiology*, 2005; 31 (3): 316–23; K. Korotkov et al., 'Assessing biophysical energy transfer mechanisms', op. cit.

30. Clair A. Francomano and Wayne B. Jonas, in Ronald A. Chez (ed.), *Proceedings: Measuring the Human Energy Field: State of the Science*. The Gerontology Research Center, National Institute of Aging, National Institutes of Health, Baltimore, Maryland, April 17–18, 2002.

31. S. Kolmakow et al., 'Gas discharge visualization technique and spectrophotometry in detection of field effects', *Mechanisms of Adaptive Behavior, Abstracts of International Symposium*, St Petersburg, 1999: 79.

32. Interview with K. Korotkov, March 2006.

Chapter 4: Hearts that Beat as One

1. All details of the Love Study were gleaned from multiple interviews with Dean Radin, Marilyn Schlitz and Jerome Stone, April 2005–June 2006.

2. F. Sicher, E. Targ et al., 'A randomized double-blind study of the effect of distant healing in a population with advanced AIDS: report of a small scale study', *Western Journal of Medicine*, 1998; 168 (6): 356–63; also multiple interviews with E. Targ, 1999–2001.

3. M. Schlitz and W. Braud, 'Distant intentionality and healing: assessing the evidence', *Alternative Therapies in Health and Medicine*, 1997; 3 (6): 62–73.

4. M. Schlitz and S. LaBerge, 'Autonomic detection of remote observation: two conceptual replications', in D. J. Bierman (ed.), *Proceedings of Presented Papers, 37th Annual Parapsychological Association Convention, Amsterdam*, Fairhaven, Mass.: Parapsychological Association, 1994: 352–60.

5. S. Schmidt et al., 'Distant intentionality and the feeling of being stared at: Two metaanalyses', *British Journal of Psychology*, 2004; 95: 235–47, as reported in D. Radin, *Entangled Minds*, New York: Paraview, 2006: 135.

6. L. Standish et al., 'Electroencephalographic evidence of correlated event-related signals between the brains of spatially and sensory isolated human subjects', *The Journal of Alternative and Complementary Medicine*, 2004; 10 (2): 307–14.

7. Radin, *Entangled Minds*, op. cit.: 136.

8. Charles Tart, 'Physiological correlates of psi cognition', *International Journal of Parapsychology*, 1963: 5; 375–86.

9. T. D. Duane and T. Behrendt, 'Extrasensory electroencephalographic induction between identical twins', *Science*, 1965; 150: 367.

10. J. Wackerman et al., 'Correlations between brain electrical activities of two spatially separated human subjects', *Neuroscience Letters*, 2003; 336: 60–4.

11. J. Grinberg-Zylberbaum et al., 'The Einstein-Podolsky-Rosen paradox in the brain: The transferred potential', *Physics Essays*, 1994; 7 (4): 422–28.

12. J. Grinberg-Zylberbaum and J. Ramos, 'Patterns of interhemisphere correlations during human communication', *International Journal of Neuroscience*, 1987; 36: 41–53; J. Grinberg-Zylberbaum et al., 'Human communication and the electrophysiological activity of the brain,' *Subtle Energies and Energy Medicine*, 1992; 3 (3): 25–43.

13. L. J. Standish et al., 'Electroencephalographic evidence of correlated event-related signals', op. cit.

14. L. J., Standish et al., 'Evidence of correlated functional magnetic resonance imaging signals between distant human brains', *Alternative Therapies in Health and Medicine*, 2003; 9 (1): 122–5; T. Richards et al., 'Replicable functional magnetic resonance imaging evidence of correlated brain signals between physically and

sensory isolated subjects', *Journal of Alternative and Complementary Medicine*, 2005; 11 (6): 955–63.

15. M. Kittenis et al., 'Distant psychophysiological interaction effects between related and unrelated participants', *Proceedings of the Parapsychological Association Convention*, 2004: 67–76, as reported in Radin, *Entangled Minds*, op. cit.: 138–9.

16. D. I. Radin, 'Event related EEG correlations between isolated human subjects', *Journal of Alternative and Complementary Medicine*, 2004; 10: 315–24.

17. M. Cade and N. Coxhead, *The Awakened Mind*, 2nd edn, Shaftesbury: Element, 1986.

18. S. Fahrion et al., 'EEG amplitude, brain mapping and synchrony in and between a bioenergy practitioner and client during healing', *Subtle Energies and Energy Medicine*, 1992; 3 (1): 19–52.

19. M. Yamamoto, 'An experiment on remote action against man in sensory shielding condition, Part 2', *Journal of the International Society of Life Information Sciences*, 1996; 14 (2): 228–39, as reported in Larry Dossey, *Be Careful What You Pray For … You Just Might Get It: What We Can Do About the Unintentional Effect of Our Thoughts, Prayers, and Wishes*, San Francisco: HarperSanFrancisco, 1998: 182–3.

20. M. Yamamoto et al., 'An experiment on remote action against man in sense shielding condition', *Journal of the International Society of Life Information Sciences*, 1996; 14 (1): 97–9.

21. D. I. Radin, 'Unconscious perception of future emotions: An experiment in presentiment', *Journal of Scientific Exploration*, 1997; 11 (2): 163–80. First presented before the annual meeting of the Parapsychological Association in August 1996. For a full description of the Radin experiment, see D. Radin, *The Conscious Universe*, London: HarperCollins, 1997: 119–24.

22. R. McCraty et al., 'Electrophysiological evidence of intuition: Part 2: A system-wide process?' *The Journal of Alternative and Complementary Medicine*, 2004; 10 (2): 325–36.

23. J. Andrew Armour and Jeffrey L. Ardell (eds.), *Basic and Clinical Neurocardiology*, Oxford: Oxford University Press, 2004.

24. R. McCraty et al., 'The electricity of touch: Detection and measurement of cardiac energy exchange between people', in Karl H. Pribram (ed.), *Brain and Values: Is a Biological Science of Values Possible?* Mahwah, NJ: Lawrence Erlbaum Associates, 1998: 359–79.

25. M. Gershon, *The Second Brain: A Groundbreaking New Understanding of Nervous Disorders of the Stomach and Intestine*, London: HarperCollins, 1999.

26. D. I. Radin and M. J. Schlitz, 'Gut feelings, intuition, and emotions: An exploratory study', *Journal of Alternative and Complementary Medicine*, 2005; 11 (5): 85–91.

27. D. Radin, 'Event-related electroencephalographic correlations between isolated human subjects', *The Journal of Alternative and Complementary Medicine*, 2004; 10 (2): 315–23.

28. Dean Radin has devoted an excellent book to the subject: see D. Radin, *Entangled Minds*, op cit.

29. J. Stone, Course Handbook: Training in Compassionate-Loving Intention, 2003; J. Stone et al., 'Effects of a compassionate/loving intention as a therapeutic intervention by partners of cancer patients: A randomized controlled feasibility study', in press.

30. M. Murphy et al., *The Physiological and Psychological Effects of Meditation: A Review of Contemporary Research with a Comprehensive Bibliography, 1931–1996*, Petaluma, Calif.: The Institute of Noetic Sciences, 1997.

31. E. P. Van Wijk et al., 'Anatomic characterization of human ultra-weak photon emission in practitioners of Transcendental Meditation™ and control subjects', *The Journal of Alternative and Complementary Medicine*, 2006; 12 (1): 31–8.

32. R. McCraty et al., 'Head-heart entrainment: A preliminary survey', in *Proceedings of the Brain-Mind Applied Neurophysiology EEG Neurofeedback Meeting*. Key West, Florida, 1996.

33. R. McCraty, 'Influence of cardiac afferent input on heart-brain synchronization and cognitive performance, Institute of HeartMath, Boulder Creek, California', *International Journal of Psychophysiology*, 2002; 45 (1–2): 72–3.

34. G. R. Schmeidler, *Parapsychology and Psychology*, Jefferson: McFarland and Company, 1988 as cited in J. Stone, Course Handbook, op. cit.; L. Dossey, *Healing Words: The Power of Prayer and the Practice of Medicine*, San Francisco: HarperSanFrancisco, 1993.

35. D. Radin et al., 'Effects of motivated distant intention on electrodermal activity.' Paper presented at the Annual Conference of the Parapsychological Association, Stockholm, Sweden, August 2006.

Chapter 5: Entering Hyperspace

1. H. Benson et al., 'Body temperature changes during the practice of g tum-mo (heat) yoga', *Nature*, 1982; 295: 234–6; H. Benson, 'Body temperature changes during the practice of g tum-mo yoga (matters arising)', *Nature*, 1982; 298: 402.

2. H. Benson et al., 'Three case reports of the metabolic and electroencephalographic changes during advanced Buddhist meditation techniques', *Behavioral Medicine*, 1990; 16 (2): 90–5.

3. The most celebrated was the Investigating the Mind conference at Massachusetts Institute of Technology, September 2005, which featured the Dalai Lama.

4. I am indebted to Stanley Krippner, who supplied me with a list of some 50 healers from a rich variety of traditions. I assembled a questionnaire, which I sent out to all 50. Some 15 replied in detail.

5. Cooperstein's study eventually was published: M. A. Cooperstein, 'The myths of healing: A summary of research into transpersonal healing experience', *Journal of the American Society for Psychical Research*, 1992; 86: 99–133. I am also indebted to him for his in-depth analysis of the commonalities between healers.

6. Information about Krippner's vast catalogue of work was also gleaned from numerous interviews between him and the author, April 2005–March 2006 and correspondence, 2005–2006.

7. S. Krippner, 'The technologies of shamanic states of consciousness', in M. Schlitz et al. (eds.), *Consciousness and Healing: Integral Approaches to Mind-Body Medicine*, St. Louis, Mo.: Elsevier Churchill Livingstone, 2005: 376–90.

8. Jilek W. G. Salish, *Indian Mental Health and Culture Change: Psychohygienic and Therapeutic Aspects of the Guardian Spirit Ceremonial*, New York: Hold Rinehart & Winston, 1974.

9. All information about Bruce Frantzis the result of various interviews, April 2005–March 2006.

10. B. K. Frantzis, *Relaxing Into Your Being: Breathing, Chi and Dissolving the Ego*, Berkeley, Calif.: North Atlantic Books, 1998.

11. Murphy, *Meditation*, op. cit.

12. W. Singer, 'Neuronal synchrony: a versatile code for the definition of relations?' *Neuron*, 1999; 24: 49–65; F. Varela et al., *Nature Reviews Neuroscience*, 2001; 2: 229–39, as reported in A. Lutz et al., 'Long-term meditators self-induce high-amplitude gamma synchrony during mental practice', *Proceedings of the National Academy of Science*, 2004; 101 (46): 16369–73.

13. O. Paulsen and T. J. Sejnowski, 'Natural patterns of activity and long-term synaptic plasticity', *Current Opinion in Neurobiology*, 2000; 10: 172–9, as reported in Lutz, 'Long-term meditators', op. cit.

14. Although the majority of studies carried out on meditation demonstrate that meditation leads to an increase in alpha rhythms (see Murphy, *Meditation*, op. cit.), the following are just a few that show that during meditation, subjects evidence spurts of high-frequency beta waves of twenty to forty cycles per second, usually during moments of intense concentration or ecstasy: J. P. Banquet, 'Spectral analysis of the EEG in meditation', *Electroencephalography and Clinical Neurophysiology*, 1973; 35: 143–51; P. Fenwick et al., 'Metabolic and EEG changes during Transcendental Meditation: An explanation', *Biological Psychology*, 1977; 5 (2): 101–18; M. A. West, 'Meditation and the EEG', *Psychological Medicine*, 1980; 10 (2): 369–75; J. C. Corby et al., 'Psychophysiological correlates of the practice of Tantric Yoga meditation', *Postgraduate Medical Journal*, 1985; 61: 301–4.

15. N. Das and H. Gastaut, 'Variations in the electrical activity of the brain, heart and skeletal muscles during yogic meditation and trance', *Electroencephalography and Clinical Neurophysiology*, 1955, Supplement no. 6: 211–19.

16. Murphy, *Meditation*, cites 10 studies showing that heart rate accelerates during these peak moments of meditation.

17. W. W. Surwillo and D. P. Hobson, 'Brain electrical activity during prayer', *Psychological Reports*, 1978; 43 (1): 135–43.

18. Murphy, *Meditation*, op. cit.

19. Lutz et al., 'Long-term meditators', op. cit.

20. Richard J. Davidson et al., 'Alterations in brain and immune function produce by mindfulness meditation', *Psychosomatic Medicine*, 2003; 65: 564–70.

21. Krippner, 'Shamanic states of consciousness', op. cit.

22. Murphy, *Meditation*, op. cit.

23. L. Bernardi et al., 'Effect of rosary prayer and yoga mantras on autonomic cardiovascular rhythms: comparative study', *British Medical Journal*, 2001; 323: 1446–9.

24. Fenwick et al., 'Metabolic and EEG changes during Transcendental Meditation', op. cit.

25. D. Goleman, *Emotional Intelligence*, London: Bloomsbury Press, 1996.

26. D. Goleman, 'Meditation and consciousness: An Asian approach to mental health', *American Journal of Psychotherapy*, 1976; 30 (1): 41–54; G. Schwartz, 'Biofeedback, self-regulation, and the patterning of physiological processes', *American Scientist*, 1975; 63 (3): 314–24; D. Goleman, 'Why the brain blocks daytime dreams', *Psychology Today*, 1976; March: 69–71.

27. P. Williams and M. West, 'EEG responses to photic stimulation in persons experienced at meditation', *Electroencephalography and Clinical Neurophysiology*,

1975; 39 (5): 519–22; B. K. Bagchi and M. A. Wenger, 'Electrophysiological correlates of some yogi exercises', *Electroencephalography and Clinical Neurophysiology*, 1957; (7): 132–49.

28. D. Brown, M. Forte and M. Dysart, 'Visual sensitivity and mindfulness meditation', *Perceptual and Motor Skills*, 1984; 58 (3): 775–84; and 'Differences in visual sensitivity among mindfulness meditators and non-meditators', *Perceptual and Motor Skills*, 1984; 58 (3): 727–33.

29. S. W. Lazar et al., 'Functional brain mapping of the relaxation response and meditation', *NeuroReport*, 2000; 11: 1581–5.

30. C. Alexander et al., 'EEG and SPECT data of a selected subject during psi tests: The discovery of a neurophysiological correlate', *Journal of Parapsychology*, 1998; 62 (2): 102–4.

31. L. LeShan, *The Medium, the Mystic and the Physicist: Towards a Theory of the Paranormal*, New York: Helios Press, 2003.

32. Cooperstein, 'The myths of healing', op. cit.

33. S. Krippner, 'Trance and the Trickster: Hypnosis as a liminal phenomenon', *International Journal of Clinical and Experimental Hypnosis*, 2005; 53 (2): 97–118.

34. E. Hartmann, *Boundaries in the Mind: A New Theory of Personality*, New York: Basic Books, 1991, as quoted in Krippner, 'Trance and the Trickster', op. cit.

35. M. J. Schlitz and Charles Honorton, 'Ganzfeld psi performance within an artistically gifted population', *Journal of the American Society for Psychical Research*, 1992; 86 (2): 83–98.

36. S. Krippner et al., 'Working with Ramtha: Is it a "high risk" procedure?' *Proceedings of Presented Papers: The Parapsychological Association 41st Annual Convention*, 1998: 50–63.

37. The various tests included the Absorption Subscale of the Differential Personality Questionnaire, the Dissociative Experiences Scale and the Boundary Questionnaire.

38. S. Krippner et al., 'The Ramtha phenomenon: Psychological, phenomenological, and geomagnetic data', *Journal of the American Society for Psychical Research*, 1998; 92: 1–24.

39. F. Sicher, E. Targ et al., 'A randomized double-blind study', op. cit.

40. Various conversations and correspondence between E. Targ and the author, October 1999–June 2001.

41. Interview with E. Targ, California, October 1999; J. Barrett, 'Going the distance', *Intuition*, 1999; June/July: 30–1.

42. D. J. Benor, *Healing Research: Holistic Energy Medicine and Spirituality*, 4 vols., Deddington, Oxfordshire: Helix Editions Ltd, 1993.

43. http://www.wholistichealingresearch.com.

44. Benor, *Healing Research*, vol. 1, op. cit.: 54–5.

45. Cooperstein, 'The myths of healing', op. cit.

46. M. Freedman et al., 'Effects of frontal lobe lesions on intentionality and random physical phenomena', *Journal of Scientific Exploration*, 2003; 17 (4): 651–68.

47. E. d'Aquili and A. Newberg, *Why God Won't Go Away: Brain Science and the Biology of Belief*, New York: Ballantine Books, 2001.

Chapter 6: In the Mood

1. All details about M. Krucoff's trip to India and decision to study prayer from interviews, August 2006.
2. R. C. Byrd, 'Positive therapeutic effects of intercessory prayer in a coronary care unit population', *Southern Medical Journal*, 1988; 81 (7): 826–9.
3. W. Harris et al., 'A randomised, controlled trial of the effects of remote, intercessory prayer on outcomes in patients admitted to the coronary care unit', *Archives of Internal Medicine*, 1999; 159 (19): 2273–8.
4. M. Krucoff, 'Integrative noetic therapies as adjuncts to percutaneous intervention during unstable coronary syndromes: Monitoring and Actualization of Noetic Training (MANTRA) feasibility pilot', *American Heart Journal*, 2001; 142 (5): 760–7.
5. M. Krucoff announced the results at the Second Conference on the Integration of Complementary Medicine into Cardiology, a meeting sponsored by the American College of Cardiology, October 14, 2003.
6. M. Krucoff et al., 'Music, imagery, touch and prayer as adjuncts to interventional cardiac care: The Monitoring and Actualisation of Noetic Trainings (MANTRA) II randomised study', *The Lancet*, 2005; 366: 211–17.
7. J. M. Aviles et al., 'Intercessory prayer and cardiovascular disease progression in a coronary care unit population: a randomized controlled trial', *Mayo Clinic Proceedings*, 2001; 76 (12): 1192–8.
8. H. Benson, *The Relaxation Response*, New York: William Morrow, 1975.
9. M. Krucoff et al., Editorial: 'From efficacy to safety concerns: A STEP forward or a step back for clinical research and intercessory prayer? The Study of Therapeutic Effects of Intercessory Prayer (STEP)', *American Heart Journal*, 2006; 151; 4: 762.
10. H. Benson et al., 'Study of the therapeutic effects of intercessory prayer (STEP) in cardiac bypass patients: A multi-center randomized trial of uncertainty and certainty of receiving intercessory prayer', *American Heart Journal*, 2006; 151 (4): 934–42.
11. Krucoff et al., 'A STEP forward', op. cit.
12. Editorial: 'MANTRA II: Measuring the unmeasurable?' *The Lancet*, 2005; 366 (9481): 178.
13. Letter to the editor, *American Heart Journal*, sent to author, 2006.
14. Krucoff et al., 'A STEP forward', op. cit.
15. B. Greyson, 'Distance healing of patients with major depression', *Journal of Scientific Exploration*, 1996; 10 (4): 447–65.
16. L. Dossey, *Meaning and Medicine: Lessons from a Doctor's Tales of Breakthough Healing*, London: Bantam, 1991; Dossey, *Healing Words*, op. cit.
17. L. Dossey, 'Prayer experiments: Science or folly? Observations on the Harvard prayer study', *Network Review (UK)*, 2006; 91: 22–3.
18. Ibid.
19. Harris, 'Effects of remote intercessory prayer', op. cit.
20. www.officeofprayerresearch.org.
21. Benor, *Healing Research*, op. cit.
22. J. Astin et al., 'The efficacy of "distant healing": A systematic review of randomized trials', *Annals of Internal Medicine*, 2000; 132: 903–10.
23. B. Rubik et al., 'In vitro effect of Reiki treatment on bacterial cultures: Role of experimental context and practitioner well-being', *The Journal of Alternative and Complementary Medicine*, 2006; 12 (1): 7–13.

24. I. R. Bell et al., 'Development and validation of a new global well-being outcomes rating scale for integrative medicine research', *BMC Complementary and Alternative Medicine*, 2004; 4: 1.

25. Ibid.

26. S. O'Laoire, 'An experimental study of the effects of distant, intercessory prayer on self-esteem, anxiety and depression', *Alternative Therapies in Health and Medicine*, 1997; 3 (6): 19–53.

27. Rubik et al., 'In vitro effect', op, cit.

28. K. Reece et al., 'Positive well-being changes associated with giving and receiving Johrei healing', *The Journal of Alternative and Complementary Medicine*, 2005; 11 (3): 455–7.

29. M. Schlitz, 'Can science study prayer?' *Shift: At the Frontiers of Consciousness*, 2006; September–November (12): 38–9.

30. Dossey, 'Prayer experiments', op. cit.

31. J. Achterberg et al., 'Evidence for correlations between distant intentionality and brain function in recipients: a functional magnetic resonance imagining analysis', *The Journal of Alternative and Complementary Medicine*, 2005; 11 (6): 965–71.

32. Ibid.

33. K. A. Wientjes, 'Mind-body techniques in wound healing', *Ostomy/Wound Management*, 2002; 48 (11): 62–7.

34. J. K. Keicolt-Glaser, 'Hostile marital interactions, proinflammatory cytokine production, and wound healing', *Archives of General Psychiatry*, 2005; 62 (12): 1377–84.

35. Krucoff, '(MANTRA) II', op. cit.

Chapter 7: The Right Time

1. For all details about Michael Persinger's experiments, interviews and correspondence with Persinger, August 2006 and a member of his neuroscientist team, Todd Murphy, May 23, 2006. Also, J. Hitt, 'This is your brain on God', *Wired*, November 1999; R. Hercz, 'The God helmet', *SATURDAYNIGHT* magazine, October 2002: 40–6; B. Raynes, 'Interview with Todd Murphy', *Alternative Perceptions Magazine* online April 2004 (No. 78), plus T. Murphy's website: www.spiritualbrain.com and M. Persinger's home page at the Laurentian University website: www.laurentian.ca/Neursci/_people/Persinger.htm.

2. Neuroscientist Todd Murphy developed this theory and successfully demonstrated its validity in Persinger's laboratory.

3. The main background of Halberg's early life is taken from F. Halberg, 'Transdisciplinary unifying implications of circadian findings in the 1950s', *Journal of Circadian Rhythms*, 2003; 1: 2.

4. G. Cornélissen et al., 'Is a birth-month-dependence of human longevity influenced by half-yearly changes in geomagnetics?' 'Physics of Auroral Phenomena', Proceedings. XXV Annual Seminar, Apatity: Polar Geophysical Institute, Kola Science Center, Russian Academy of Science, February 26–March 1, 2002: 161–6; A. M. Vaiserman et al., 'Human longevity: related to date of birth?' Abstract 9, 2nd International Symposium: Workshop on Chronoastrobiology and Chronotherapy, Tokyo Kasei University, Tokyo, Japan, November 2001.

5. O. N. Larina et al., 'Effects of spaceflight factors on recombinant protein expression in *E.coli* producing strains', in 'Biomedical Research on the Science/NASA

Project', Abstracts of the Third US/Russian Symposium, Huntsville, Alabama, November 10–13, 1997: 110–11.

6. D. Hillman et al., 'About-10 yearly (circadecennian) cosmo-helio geomagnetic signatures in *Acetabularia*', *Scripta Medica* (BRNO), 2002; 75 (6): 303–8.

7. P. A. Kashulin et al., 'Phenolic biochemical pathway in plants can be used for the bioindication of heliogeophysical factors', 'Physics of Auroral Phenomena', Proceedings. XXV Annual Seminar, Apatity: Polar Geophysical Institute, Kola Science Center, Russian Academy of Science, February 26–March 1, 2002: 153–6.

8. V. M. Petro et al., 'An influence of changes of magnetic field of the Earth on the functional state of humans in the conditions of space mission', Proceedings, International Symposium 'Computer Electro-Cardiograph on Boundary of Centuries', Moscow, Russian Federation, 27–30 April, 1999.

9. K. F. Novikova and B. A. Ryvkin, 'Solar activity and cardiovascular diseases', in M. N. Gnevyshev and A. I. Ol (eds.), *Effects of Solar Activity on the Earth's Atmosphere and Biosphere*, Academy of Science, USSR (translated from the Russian), Jerusalem: Israel Program for Scientific Translations, 1977: 184–200.

10. G. Cornélissen et al., 'Chronomes, time structures, for chronobioengineering for "a full life"', *Biomedical Instrumentation and Technology*, 1999; 33 (2): 152–87.

11. V. N. Oraevskii et al., 'Medico-biological effect of natural electromagnetic variations', *Biofizika*, 1998; 43 (5): 844–8; V. N. Oraevskii et al., 'An influence of geomagnetic activity on the functional status of the body', *Biofizika*, 1998; 43 (5): 819–26.

12. I. Gurfinkel et al., 'Assessment of the effect of a geomagnetic storm on the frequency of appearance of acute cardiovascular pathology', *Biofizika*, 1998; 43 (4): 654–8; J. Sitar, 'The causality of lunar changes on cardiovascular mortality', *Casopis Lekaru Ceskych*, 1990; 129: 1425–30.

13. F. Halberg et al., 'Cross-spectrally coherent about 10-5- and 21-year biological and physical cycles, magnetic storms and myocardial infarctions', *Neuroendrocrinology Letters*, 2000; 21: 233–58.

14. M. N. Gnevyshev, 'Essential features of the 11-year solar cycle', *Solar Physics*, 1977; 51: 175–82.

15. G. Cornélissen et al., 'Non-photic solar associations of heart rate variability and myocardial infarction', *Journal of Atmospheric and Solar-terrestrial Physics*, 2002; 64: 707–20.

16. A. R. Allahverdiyev et al., 'Possible space weather influence on functional activity of the human brain', Proceedings, Space Weather Workshop: Looking Towards a European Space Weather Programme, December 17–19, 2001.

17. E. Babayev, 'Some results of investigations on the space weather influence on functioning of several engineering-technical and communication systems and human health', *Astronomical and Astrophysical Transactions*, 2003; 22 (6): 861–7; G. Y. Mizon and P. G. Mizun, *Space and Health*, Moscow: 'Znanie', 1984.

18. E. Stoupel, 'Relationship between suicide and myocardial infarction with regard to changing physical environmental conditions', *International Journal of Biometeorology*, 1994; 38 (4): 199–203; E. Stoupel et al., 'Clinical cosmobiology: the Lithuanian study, 1990–1992', *International Journal of Biometeorology*, 1995; 38: 204–8; E. Stoupel et al., 'Suicide-homicide temporal interrelationship, links with other fatalities and environmental physical activity', *Crisis*, 2005; 26: 85–9.

19. Avi Raps et al., 'Geophysical Variables and Behavior: LXIX. Solar activity and admission of psychiatric inpatients', *Perceptual and Motor Skills*, 1992; 74: 449; H. Friedman et al., 'Geomagnetic parameters and psychiatric hospital admissions', *Nature*, 1963; 200: 626–8.

20. M. Mikulecky, 'Lunisolar tidal waves, geomagnetic activity and epilepsy in the light of multivariate coherence', *Brazilian Journal of Medicine*, 1996; 29 (8): 1069–72; E. A. McGugan, 'Sudden unexpected deaths in epileptics – a literature review', *Scottish Medical Journal*, 1999; 44 (5): 137–9.

21. A. Michon et al., 'Attempts to simulate the association between geomagnetic activity and spontaneous seizures in rats using experimentally generated magnetic fields', *Perceptual and Motor Skills*, 1996; 82 (2): 619–26; Y. Bureau and M. Persinger, 'Geomagnetic activity and enhanced mortality in rats with acute (epileptic) limbic lability', *International Journal of Biometeorology*, 1992; 36: 226–32.

22. Y. Bureau and M. Persinger, 'Decreased latencies for limbic seizures induced in rats by lithium-pilocarpine occur when daily average geomagnetic activity exceeds 20 nanotesla', *Neuroscience Letters*, 1995; 192: 142–4; A. Michon and M. A. Persinger, 'Experimental simulation of the effects of increased geomagnetic activity upon nocturnal seizures in epileptic rats', *Neuroscience Letters*, 1997; 224: 53–6.

23. M. Persinger, 'Sudden unexpected death in epileptics following sudden, intense, increases in geomagnetic activity: Prevalence of effect and potential mechanisms', *International Journal of Biometeorology*, 1995; 38: 180–7; R. P. O'Connor and M. A. Persinger, 'Geophysical Variables and Behavior: LXXXII. A strong association between sudden infant death syndrome (SIDS) and increments of global geomagnetic activity – possible support for the melatonin hypothesis', *Perceptual and Motor Skills*, 1997; 84: 395–402.

24. B. McKay and M. Persinger, 'Geophysical Variables and Behavior: LXXXVII. Effects of synthetic and natural geomagnetic patterns on maze learning', *Perceptual and Motor Skills*, 1999; 89 (3 pt 1): 1023–4

25. Radin, *Conscious Universe*, op. cit.

26. D. Radin, 'Evidence for relationship between geomagnetic field fluctuations and skilled physical performance.' Presentation made at the 11th Annual Meeting of the Society for Scientific Exploration, Princeton, New Jersey, June 1992.

27. S. W. Tromp, *Biometeorology*, London: Heyden, 1980.

28. I. Stoilova and T. Zdravev, 'Influence of the geomagnetic activity on the human functional systems', *Journal of the Balkan Geophysical Society*, 2000; 3 (4): 73–6.

29. J. S. Derr and M. A. Persinger, 'Geophysical Variables and Behavior: LIV. Zeitoun (Egypt) apparitions of the Virgin Mary as tectonic strain-induced luminosities', *Perceptual and Motor Skills*, 1989; 68: 123–8.

30. M. A. Persinger and S. A. Koren, 'Experiences of spiritual visitation and impregnation: potential induction by frequency-modulated transients from an adjacent clock', *Perceptual and Motor Skills*, 2001; 92 (1): 35–6.

31. M. A. Persinger et al., 'Differential entrainment of electroencephalographic activity by weak complex electromagnetic fields', *Perceptual and Motor Skills*, 1997; 84 (2): 527–36.

32. M. A. Persinger, 'Increased emergence of alpha activity over the left but not the right temporal lobe within a dark acoustic chamber: Differential response of the left

but not the right hemisphere to transcerebral magnetic fields', *International Journal of Psychophysiology*, 1999; 34 (2): 163–9.

33. Interview with Todd Murphy, May 23, 2006.

34. W. G. Braud and S. P. Dennis, 'Geophysical Variables and Behavior: LVIII. Autonomic activity, hemolysis and biological psychokinesis: Possible relationships with geomagnetic field activity', *Perceptual and Motor Skills*, 1989; 68: 1243–54.

35. Ibid.

36. McTaggart, *The Field*, op. cit.: 167–8.

37. M. A. Persinger and S. Krippner, 'Dream ESP experiments and geomagnetic activity', *Journal of the American Society for Psychical Research*, 1989; 83: 101–16; S. Krippner and M. Persinger, 'Evidence for enhanced congruence between dreams and distant target material during periods of decreased geomagnetic activity', *Journal of Scientific Exploration*, 1996; 10, (4): 487–93.

38. M. Ullman et al., *Dream Telepathy: Experiments in ESP*, Jefferson: McFarland, 1989.

39. Ibid.

40. M. A. Persinger, 'ELF field meditation in spontaneous psi events. Direct information transfer or conditioned elicitation?' *Psychoenergetic Systems*, 1975; 3: 155–69; M. A. Persinger, 'Geophysical Variables and Behavior: XXX. Intense paranormal activities occur during days of quiet global geomagnetic activity', *Perceptual and Motor Skills*, 1985; 61: 320–2.

41. M. H. Adams, 'Variability in remote-viewing performance: Possible relationship to the geomagnetic field', in D. H. Weiner and D. I. Radin (eds.), *Research in Parapsychology*, Metuchen, NJ: Scarecrow Press, 1986: 25. [cf n.19, ch.8]

42. J. N. Booth et al., 'Ranking of stimuli that evoked memories in significant others after exposure to circumcerebral magnetic fields: Correlations with ambient geomagnetic activity', *Perceptual and Motor Skills*, 2002; 95(2): 555–8.

43. M. A. Persinger et al., 'Differential entrainment of electroencephalographic activity by weak complex electromagnetic fields', *Perceptual and Motor Skills*, 1997; 84 (2): 527–36.

44. M. A. Persinger, 'Enhancement of images of possible memories of others during exposure to circumcerebral magnetic fields: Correlations with ambient geomagnetic activity', *Perceptual and Motor Skills*, 2002; 95 (2): 531–43.

45. S. A. Koren and M. A Persinger, 'Possible disruption of remote viewing by complex weak magnetic fields around the stimulus site and the possibility of accessing real phase space: A pilot study', *Perceptual and Motor Skills*, 2002; 95 (3 Pt 1): 989–98.

46. S. Krippner, 'Possible geomagnetic field effects in psi phenomena.' Paper presented at international parapsychology conference in Recife, Brazil, November 1997.

47. Braud and Dennis, 'Geophysical Variables and Behavior: LVIII', op. cit.

48. S. J. P. Spottiswoode, 'Apparent association between effect size in free response anomalous cognition experiments and local sidereal time', *Journal of Scientific Exploration*, 1997; 11 (2): 109–22.

49. S. J. P. Spottiswoode and E. May, 'Evidence that free response anomalous cognitive performance depends upon local sidereal time and geomagnetic fluctuations', Presentation Abstracts, Sixteenth Annual Meeting of the Society for Scientific Exploration, June 1997: 8.

50. A. P. Krueger and D. S. Sobel, 'Air ions and health', in David S. Sobel (ed.), *Ways of Health: Holistic Approaches to Ancient and Contemporary Medicine*, New York: Harcourt Brace Jovanovich, 1979.

Chapter 8: The Right Place

1. William Tiller's major books on crystallization include: *An Introduction to Computer Simulation in Applied Science,* New York: Plenum, 1992: *The Science of Crystallization: Microscopic Interfacial Phenomena,* Cambridge: Cambridge University Press, 1991: *The Science of Crystallization: Macroscopic Phenomena and Defect Generation,* Cambridge: Cambridge University Press, 1992.

2. All personal details about William Tiller have resulted from multiple interviews, April 2005–January 2006.

3. O. Warburg, *New Methods of Cell Physiology Applied to Cancer and Mechanism of X-ray Action,* New York: John Wiley and Sons, 1962, as quoted in W. Tiller et al., *Conscious Acts of Creation: The Emergency of a New Physics,* Walnut Creek, Calif.: Pavior Publishing, 2001: 144–6. All description of experiment derived from interview with Dr Tiller, Boulder, Colorado, April 29, 2005, plus information from *Conscious Acts* and W. Tiller et al., *Some Science Adventures with Real Magic,* Walnut Creek, Calif.: Pavior Publishing, 2005.

4. M. J. Kohane, 'Energy, development and fitness in *Drosophila melanogaster*', *Proceedings of the Royal Society* (B), 1994; 257: 185–91, in Tiller et al., *Conscious Acts,* op. cit.: 147.

5. William A. Tiller and Walter E. Dibble, Jr., 'New experimental data revealing an unexpected dimension to materials science and engineering', *Material Research Innovation,* 2001; 5: 21–34.

6. Tiller and Dibble, 'New experimental data', op. cit.

7. Ibid.

8. Ibid.

9. Tiller et al., *Conscious Acts,* op. cit.: 180.

10. Tiller et al., *Conscious Acts,* op. cit.: 175.

11. Tiller et al., *Conscious Acts,* op. cit.: 216.

12. H. Pagels, *The Cosmic Code,* New York: Simon and Schuster, 1982.

13. Tiller et al., *Conscious Acts,* op. cit.: 216.

14. Tiller et al., *Science Adventures,* op. cit.: 34.

15. Interview with W. Tiller, April 2005.

16. Tiller et al., *Conscious Acts,* op. cit.: 182.

17. Correspondence between Tiller and Michael Kohane, 2005.

18. Tiller and Dibble, 'New experimental data', op. cit.

19. G. K. Watkins and A. M. Watkins, 'Possible PK influence on the resuscitation of anesthetized mice', *Journal of Parapsychology,* 1971; 35: 257–72; G. K. Watkins et al., 'Further studies on the resuscitation of anesthetized mice', in W. G. Roll, R. L. Morris and J. Morris (eds.), *Research in Parapsychology,* Metuchen, NJ: Scarecrow Press, 1973: 157–9.

20. R. Wells and J. Klein, 'A replication of a "psychic healing" paradigm', *Journal of Parapsychology,* 1972; 36: 144–9.

21. See McTaggart, *The Field,* op. cit.: 205–7.

22. D. Radin, 'Beyond belief: Exploring interaction among body and environment', *Subtle Energies and Energy Medicine,* 1992; 2 (3): 1–40; D. Radin, 'Environmental modulation and statistical equilibrium in mind-matter interaction', *Subtle Energies and Energy Medicine,* 1993; 4 (1): 1–30.

23. D. Radin et al., 'Effects of healing intention on cultured cells and truly random events', *The Journal of Alternative and Complementary Medicine,* 2004; 10: 103–12.

24. L. P. Semikhina and V. P. Kiselev, 'Effect of weak magnetic fields on the properties of water and ice', *Zabedenii, Fizika*, 1988; 5: 13–17; S. Sasaki et al., 'Changes of water conductivity induced by non-inductive coil', *Society for Mind-Body Science*, 1992; 1: 23; Tiller et al., *Conscious Acts*, op. cit.: 62.

Chapter 9: Mental Blueprints

1. All description of Ali's fighting techniques from N. Mailer, *The Fight*, London and New York: Penguin, 2000.

2. Ibid.

3. A. Richardson, 'Mental practice: A review and discussion, Part I', *Research Quarterly*, 1967; 38: 95–107; A. Richardson, 'Mental practice: A review and discussion. Part II', *Research Quarterly*, 1967; 38: 264–73.

4. J. Salmon et al., 'The use of imagery by soccer players', *Journal of Applied Sport Psychology*, 1994; 6: 116–33.

5. A. Paivio, *Mental Representations: A Dual Coding Approach*, New York and London: Oxford University Press, 1986.

6. B. S. Rushall and L. G. Lippman, 'The role of imagery in physical performance', *International Journal for Sport Psychology*, 1997; 29: 57–72.

7. A. Paivio, 'Cognitive and motivational functions of imagery in human performance', *Canadian Journal of Applied Sport Sciences*, 1985; 10 (4): 22S–28S.

8. K. E. Hinshaw, 'The effects of mental practice on motor skill performance: Critical evaluation and meta-analysis', *Imagination, Cognition and Personality*, 1991–2; 11: 3–35.

9. J. A. Swets and R. A. Bjork, 'Enhancing human performance: An evaluation of "New Age" techniques considered by the U.S. Army', *Psychological Science*, 1990; 1: 85–96; D. L. Feltz et al., 'A revised meta-analysis of the mental practice literature on motor skill learning', in D. Druckman and J. A. Swets (eds.), *Enhancing Human Performance: Issues, Theories, and Techniques*, Washington, DC: National Academy Press, 1988: 274.

10. R. J. Rotella et al., 'Cognitions and coping strategies of elite skiers: an exploratory study of young developing athletes', *Journal of Sport Psychology*, 1980; 2: 350–4.

11. R. S. Burhans et al., 'Mental imagery training: effects on running speed performance', *International Journal of Sport Psychology*, 1988; 19: 26–37.

12. B. S. Rushall, 'Covert modeling as a procedure for altering an elite athlete's psychological state', *Sport Psychologist*, 1988; 2: 131–40; B. S. Rushall, 'The restoration of performance capacity by cognitive restructuring and covert positive reinforcement in an elite athlete', in J. R. Cautela and A. J. Kearney (eds.), *Covert Conditioning Casebook*. Boston, Mass.: Thomson Brooks/Cole, 1993.

13. M. Denis, 'Visual imagery and the use of mental practice in the development of motor skills', *Canadian Journal of Applied Sport Sciences*, 1985; 10: 4S–16S.

14. Paivio, 'Cognitive and motivational functions of imagery', op. cit.

15. J. R. Cautela and A. J. Kearney (eds.), *Covert Conditioning Casebook*. Boston, Mass.: Thomson Brooks/Cole, 1993: 30–1.

16. B. Mumford and C. Hall, 'The effects of internal and external imagery on performing figures in figure skating', *Canadian Journal of Applied Sport Sciences*, 1985; 10: 171–7.

17. K. Barr and C. Hall, 'The use of imagery by rowers', *International Journal of Sport Psychology*, 1992; 23: 243–61.

18. S. C. Minas, 'Mental practice of a complex perceptual-motor skill', *Journal of Human Movement Studies*, 1978; 4: 102–7.

19. R. Bleier, *Fighting Back*, New York: Stein and Day, 1975.

20. R. L. Wilkes and J. J. Summers, 'Cognitions, mediating variables and strength performance', *Journal of Sport Psychology*, 1984; 6: 351–9.

21. R. S. Weinberg et al., 'Effects of visuo-motor behavior rehearsal, relaxation, and imagery on karate performance', *Journal of Sport Psychology*, 1981; 3: 228–38.

22. Cautela and Kearney, *Covert Conditioning*, op. cit.

23. J. Pates et al., 'The effects of hypnosis on flow states and three-point shooting in basketball players', *The Sport Psychologist*, 2002; 16: 34–47; J. Pates and I. Maynard, 'Effects of hypnosis on flow states and golf performance', *Perceptual and Motor Skills*, 2000; 9: 1057–75.

24. R. M. Suinn, 'Imagery rehearsal applications to performance enhancement', *The Behavior Therapist*, 1985; 8: 155–9.

25. L. Baroga, 'Influence on the sporting result of the concentration of attention process and time taken in the case of weight lifters', in *Proceedings of the 3rd World Congress of the International Society of Sports Psychology*, Volume 3. Madrid, Spain: Instituto Nacional de Educacion Fisica Y Deportes, 1973.

26. A. Fujita, 'An experimental study on the theoretical basis of mental training', in *Proceedings of the 3rd World Congress of the International Society of Sports Psychology*, Volume Abstracts. Madrid, Spain: Instituto Nacional de Educacion Fisica Y Deportes, 1973: 37–8.

27. Ibid.

28. Rushall and Lippman, 'The role of imagery in physical performance', op. cit.

29. G. H. Van Gyn et al., 'Imagery as a method of enhancing transfer from training to performance', *Journal of Sport and Exercise Science*, 1990; 12: 366–75.

30. G. H. Yue and K. J. Cole, 'Strength increases from the motor program; Comparison of training with maximal voluntary and imagined muscle contractions', *Journal of Neurophysiology*, 1992; 67: 114–23; V. K. Ranganathan et al., 'Increasing muscle strength by training the central nervous system without physical exercise', *Society for Neuroscience Abstracts*, 2001; 31: 17; V. K. Ranganathan et al., 'Level of mental effort determines training-induced strength increases', *Society of Neuroscience Abstracts*, 2002; 32: 768; P. Cohen, 'Mental gymnastics', *New Scientist*, November 24, 2001; 172 (2318): 17.

31. D. Smith et al., 'The effect of mental practice on muscle strength and EMG activity', *Proceedings of the British Psychological Society* annual conference, 1998; 6 (2): 116.

32. T. X. Barber, 'Changing "unchangeable" bodily processes by (hypnotic) suggestions: A new look at hypnosis, cognitions, imagining and the mind-body problem', in A. A. Sheikh (ed.), *Imagination and Healing*, Farmingdale, NY: Baywood Publishing Co., 1984. Also published in *Advances*, Spring 1984.

33. F. M. Luskin et al., 'A review of mind-body therapies in the treatment of cardiovascular disease, Part 1: Implications for the elderly', *Alternative Therapies in Health and Medicine*, 1998; 4 (3): 46–61.

34. F. M. Luskin et al., 'A review of mind/body therapies in the treatment of musculoskeletal disorders with implications for the elderly', *Alternative Therapies in Health and Medicine*, 2000; 6 (2): 46–56.

35. V. A. Hadhazy et al., 'Mind-body therapies for the treatment of fibromyalgia. A systematic review', *Journal of Rheumatology*, 2000; 27 (12): 2911–18.

36. J. A. Astin et al., 'Mind-body medicine: State of the science: Implications for practice', *Journal of the American Board of Family Practitioners*, 2003; 16 (2): 131–47.

37. J. A. Astin, 'Mind-body therapies for the management of pain', *Clinical Journal of Pain*, 2004; 20 (1): 27–32.

38. L. S. Eller, 'Guided imagery interventions for symptom management', *Annual Review of Nursing Research*, 1999; 17, 57–84.

39. J. Achterberg and G. F. Lawlis, *Bridges of the Bodymind: Behavioral Approaches for Health Care*, Champaign, Ill.: Institute for Personality and Ability Testing, 1980.

40. N. E. Miller and L. DiCara, 'Instrumental learning of heart rate changes in curarized rats: Shaping and specificity to discriminative stimulus', *Journal of Comparative and Physiological Psychology*, 1967; 63: 12–19; N. E. Miller, 'Learning of visceral and glandular responses', *Science*, 1969; 163: 434–45.

41. J.V. Basmajian, *Muscles Alive: Their Functions Revealed by Electromyography*. Baltimore, Md.: Williams and Wilkins, 1967.

42. E. Green, 'Feedback technique for deep relaxation', *Psychophysiology*, 1969; 6 (3): 371–7; E. Green et al., 'Self-regulation of internal states', in J. Rose (ed.), *Progress of Cybernetics: Proceedings of the First International Congress of Cybernetics, London, September 1969*. London: Gordon and Breach Science Publishers, 1970: 1299–318; E. Green et al., 'Voluntary control of internal states: Psychological and physiological', *Journal of Transpersonal Psychology*, 1970; 2: 1–26; D. Satinsky, 'Biofeedback treatment for headache: A two-year follow-up study', *American Journal of Clinical Biofeedback*, 1981; 4 (1): 62–5; B.V. Silver et al., 'Temperature biofeedback and relaxation training in the treatment of migraine headaches: One-year follow-up', *Biofeedback and Self Regulation*, 1979; 4 (4): 359–66.

43. B. M. Kappes, 'Sequence effects of relaxation training, EMG, and temperature biofeedback on anxiety, symptom report, and self-concept', *Journal of Clinical Psychology*, 1983; 39 (2): 203–8; G. Rose et al., 'The behavioral treatment of Raynaud's disease: A review', *Biofeedback and Self Regulation*, 1987; 12 (4): 257–72.

44. W. T. Tsushima, 'Treatment of phantom limb pain with EMG and temperature biofeedback: A case study', *American Journal of Clinical Biofeedback*, 1982; 5 (2): 150–3.

45. T. G. Dobie, 'A comparison of two methods of training resistance to visually-induced motion sickness.' Paper presented at VII International Man in Space Symposium: Physiologic adaptation of man in space, Houston, Texas, 1986. *Aviation, Space, and Environmental Medicine*, 1987; 58 (9) Sect. 2: 34–41.

46. A. Ikemi et al., 'Thermographical analysis of the warmth of the hands during the practice of self-regulation method', *Psychotherapy and Psychosomatics*, 1988; 50 (1): 22–8.

47. J. L. Claghorn, 'Directional effects of skin temperature self-regulation on regional cerebral blood flow in normal subjects and migraine patients', *American Journal of Psychiatry*, 1981; 138 (9): 1182–7.

48. M. Davis et al., *The Relaxation and Stress Reduction Workbook*, 5th edn, Oakland, Calif.: New Harbinge, 2000: 83–90.

49. J. K. Lashley et al., 'An empirical account of temperature biofeedback applied in groups', *Psychological Reports*, 1987; 60 (2): 379–88; S. Fahrion et al., 'Biobehavioral

treatment of essential hypertension: A group outcome study', *Biofeedback and Self Regulation*, 1986; 11 (4): 257–77.

50. J. Panksepp, 'The anatomy of emotions', in R. Plutchik (ed.), *Emotion: Theory, Research and Experience Vol. III. Biological Foundations of Emotions*, New York: Academic Press, 1986: 91–124.

51. J. Panksepp, 'The neurobiology of emotions: Of animal brains and human feelings', in T. Manstead and H. Wagner (eds.), *Handbook of Psychophysiology*, Chichester: John Wiley & Sons, 1989: 5–26.

52. C. D. Clemente et al., 'Postreinforcement EEG synchronization during alimentary behavior', *Electroencephalography and Clinical Neurophysiology*, 1964; 16: 335–65; M. H. Chase et al., 'Afferent vagal stimulation: Neurographic correlates of induced EEG synchronization and desynchronization', *Brain Research*, 1967; 5: 236–49.

53. M. B. Sterman, 'Neurophysiological and clinical studies of sensorimotor EEG biofeedback training: Some effects on epilepsy', *Seminars in Psychiatry*, 1973; 5 (4): 507–25; M. B. Sterman, 'Neurophysiological and clinical studies of sensorimotor EEG biofeedback training: Some effects on epilepsy', in L. Birk (ed.), *Biofeedback: Behavioral Medicine*. New York: Grune and Stratton, 1973: 147–65; M. B. Sterman, 'Epilepsy and its treatment with EEG feedback therapy', *Annals of Behavioral Medicine*, 1986; 8: 21–5; M. B. Sterman, 'The challenge of EEG biofeedback in the treatment of epilepsy: A view from the trenches', *Biofeedback*, 1997; 25 (1): 6–7; M. B. Sterman, 'Basic concepts and clinical findings in the treatment of seizure disorders with EEG operant conditioning', *Clinical Electroencephalography*, 2000; 31 (1): 45–55.

54. E. Peniston and P. J. Kulkosky, 'Alpha-theta brainwave training and beta-endorphin levels in alcoholics', *Alcoholism: Clinical and Experimental Research*, 1989; 13: 271–9; E. Peniston and P. J. Kulkosky, 'Alcoholic personality and alpha-theta brainwave training', *Medical Psychotherapy*, 1990; 3: 37–55.

55. J. Kamiya, 'Operant control of the EEG alpha rhythm', in C. Tart (ed.), *Altered States of Consciousness*, New York: John Wiley & Sons, 1969, J. Kamiya, 'Conscious control of brain waves', *Psychology Today*, April 1968: 7.

56. N. E. Schoenberger et al., 'Flexyx neurotherapy system in the treatment of traumatic brain injury: An initial evaluation', *Journal of Head Trauma Rehabilitation*, 2001; 16 (3): 260–74.

57. C. B. Kidd, 'Congenital ichthyosiform erythroderma treated by hypnosis', *British Journal of Dermatology*, 1966; 78: 101–5, as cited in Barber, 'Changing "unchangeable" bodily processes', op. cit.

58. H. Bennett, 'Behavioral anesthesia', *Advances*, 1985; 2 (4): 11–21, as reported in H. Dienstfrey, 'Mind and mindlessness in mind-body research', in M. Schlitz et al., *Consciousness and Healing: Integral Approaches to Mind-Body Healing*, St Louis, Mo.: Elsevier Churchill Livingstone, 2005: 56.

59. H. Dienstfrey, 'Mind and mindlessnes', op cit.: 51–60.

60. Dr Angel Escudero was featured on the BBC's *Your Life in Their Hands* series, May 1991. In the film, Escudero made incisions, sawed, drilled and hammered in order to break and reset the deformed leg of his fully conscious patient using his 'Noesitherapy' technique of pain control.

61. S. M. Kosslyn et al., 'Hypnotic visual illusion alters color processing in the brain', *American Journal of Psychiatry*, 2000; 157: 1279–84; Mark Henderson, 'Hypnosis really does turn black into white', *The Times*, 18 February 2002.

62. S. H. Simpson et al., 'A meta-analysis of the association between adherence to drug therapy and mortality', *British Medical Journal*, 2006; 333: 15–19.

63. Raúl de la Fuente-Fernández et al., 'Expectation and dopamine release: Mechanism of the placebo effect in Parkinson's disease', *Science*, 2001; 293 (5532): 1164–6.

64. J. B. Moseley et al., 'A controlled trial of arthroscopic surgery for osteoarthritis of the knee', *New England Journal of Medicine*, 2002; 347: 81–8.

65. S. Krippner, 'Stigmatic phenomenon: An alleged case in Brazil', *Journal of Scientific Exploration*, 2002; 16 (2): 207–24.

66. L. F. Early and J. E. Kifschutz, 'A case of stigmata', *Archives of General Psychiatry*, 1974; 30: 197–200.

67. T. Harrison, *Stigmatia: A Medieval Mystery in a Modern Age*, New York: St Martin's Press, 1994, as referenced in S. Krippner, 'Stigmatic phenomenon', op. cit.

68. B. O'Regan and Caryle Hirshberg, *Spontaneous Remission: An Annotated Bibliography*, Petaluma, Calif.: Institute of Noetic Sciences, 1993.

69. Ibid.

70. L. L. LeShan and M. L. Gassmann, 'Some observations on psychotherapy with patients with neoplastic disease', *American Journal of Psychotherapy*, 1958; 12: 723.

71. D. C. Ban Baalen et al., 'Psychosocial correlates of "spontaneous" regression of cancer', *Humane Medicine*, April 1987.

72. R. T. D. Oliver, 'Surveillance as a possible option for management of metastic renal cell carcinoma', *Seminars in Urology*, 1989; 7: 149–52.

73. P. C. Raud, 'Psychospiritual dimensions of extraordinary survival', *Journal of Humanistic Psychology*, 1989; 29: 59–83.

74. McTaggart, *The Field*, op. cit.: 132.

75. W. Braud and M. Schlitz, 'Psychokinetic influence on electrodermal activity', *Journal of Parapsychology*, 1983; 47 (2): 95–119.

76. Interview with William Braud, October, 1999.

77. Benor, *Healing Research*, op. cit.

78. S. M. Roney-Dougal and J. Solfvin, 'Field study of an enhancement effect on lettuce seeds – Replication study', *Journal of Parapsychology*, 2003; 67 (2): 279–98.

79. Dr Larry Dossey calls negative diagnoses 'medical hexing', and there is anecdotal evidence that patients often live up to their doctor's gloomy prognosis, even when there is no physical evidence that they should do so. For a potent example see the story of a leukaemia patient who was thriving until he happened to find out what he had. He was dead within a week once his illness had the label of a potentially terminal illness: L. McTaggart, *What Doctors Don't Tell You*, London: HarperCollins, 2005: 343.

Chapter 10: The Voodoo Effect

1. R. A. Blasband and Gottfried Martin, 'Biophoton emission in "orgone energy" treated cress seeds, seedlings and *Acetabularia*', International Consciousness Research Laborary, ICRL Report No 93.6.

2. Dossey, *Be Careful What You Pray For*, op. cit.: 171–2.

3. Ibid.

4. Benor, *Healing Research*, op. cit.: 261.

5. C. O. Simonton et al., *Getting Well Again*, New York: Bantam, 1980; B. Siegel, *Love, Medicine and Miracles: Lessons Learned about Self-Healing from a Surgeon's Experience with Exceptional Patients*, London: HarperCollins, 1990; A. Meares, *The Wealth Within: Self-Help Through a System of Relaxing Meditation*, Melbourne, Australia: Hill of Content, 1990.

6. For much of the research detailed in this chapter, I am especially indebted to Larry Dossey and Daniel Benor, who have detailed many of these early studies in their respective books, Dossey's *Be Careful What You Pray For … You Just Might Get It* and Benor's *Healing Research, Spiritual Healing* and his outstanding, comprehensive website: www.wholistichealingresearch.com.

7. Benor, *Healing Research*, op. cit.: 264.

8. J. Barry, 'General and comparative study of the psychokinetic effect on a fungus culture', *Journal of Parapsychology*, 1968; 32 (94): 237–43.

9. W. H. Tedder and M. L. Monty, 'Exploration of a long-distance PK: A conceptual replication of the influence on a biological system', in W. G. Roll et al. (eds.), *Research in Parapsychology*, Metuchen, NJ: Scarecrow Press, 1981: 90–3. Also see Dossey, *Be Careful What You Pray For*, op. cit.: 169; Benor, *Healing Research*, op. cit.: 268–9.

10. C. B. Nash, 'Test of psychokinetic control of bacterial mutation', *Journal of the American Society for Psychical Research*, 1984; 78: 145–52.

11. Kmetz's study was described in W. Braud et al., 'Experiments with Matthew Manning', *Journal of the Society for Psychical Research*, 1979; 50: 199–223. While the study was promising, in his review of it in *Healing Research*, Benor noted the lack of sufficient detail.

12. Dossey, *Be Careful What You Pray For*, op. cit.: 175–6.

13. Many researchers of alternative medicine maintain the same concerns about studies of Chinese medicine carried out in China. These concerns don't disregard the strong anecdotal evidence about the effectiveness of Traditional Chinese Medicine, only the scientific method of studies of its effectiveness.

14. S. Sun and C. Tao, 'Biological effect of emitted *qi* with tradescantic paludosa micronuclear technique', First World Conference for Academic Exchange of Medical Qigong. Beijing, China, 1988: 61E.

15. Ibid.

16. Dossey, *Be Careful What You Pray For*, op. cit.: 176.

17. D. J. Muehsam et al., 'Effects of Qigong on cell-free myosin phosphorylation: Preliminary experiments', *Subtle Energies and Energy Medicine*, 1994; 5 (1): 93–108, also reported in Dossey, *Be Careful What You Pray For*, op. cit.: 177–8.

18. Ibid.

19. Benor, *Healing Research*, op. cit.: 253.

20. G. Rein, *Quantum Biology: Healing with Subtle Energy*, Palo Alto, Calif.: Quantum Biology Research Labs, 1992; as reported in Benor, *Healing Research*, op. cit.: 350–2.

21. B. Grad, 'The "laying on of hands": Implications for psychotherapy, gentling and the placebo effect', *Journal of the Society for Psychical Research*, 1967; 61 (4): 286–305.

22. C. B. Nash and C. S. Nash, 'The effect of paranormally conditioned solution on yeast fermentation', *Journal of Parapsychology*, 1967; 31: 314.

23. Radin, *The Conscious Universe*, op. cit: 130.

24. An entire chapter is devoted to Jacques Benveniste in McTaggart, *The Field*, op. cit.: 59.
25. Description of these results from a telephone conversation with Jacques Benveniste, November 10, 2000.
26. J. M. Rebman et al., 'Remote influence of the autonomic nervous system by focused intention', *Subtle Energies and Energy Medicine*, 1996; 6: 111–34.
27. W. Braud and M. Schlitz, 'A method for the objective study of transpersonal imagery', *Journal of Scientific Exploration*, 1989; 3 (1): 43–63; W. Braud et al., 'Further studies of the bio-PK effect: Feedback, blocking specificity/generality', in R. White and J. Solfvin (eds.), *Research in Parapsychology*, Metuchen, NJ: Scarecrow Press, 1984: 45–8.
28. C. Watt et al., 'Exploring the limits of direct mental influence: Two studies comparing "blocking" and "co-operating" strategies', *Journal of Scientific Exploration*, 1999; 13 (3): 515–35.
29. J. Diamond, *Your Body Doesn't Lie*, New York: HarperCollins, 1979.
30. J. Diamond, *Life Energy*, New South Wales: Angus & Robertson, 1992: 71.

Chapter 11: Praying for Yesterday

1. L. Leibovici, 'Effects of remote, retroactive intercessory prayer on outcomes in patients with blood stream infection: Randomized controlled trial', *British Medical Journal*, 2001; 323 (7327): 1450–1.
2. S. Andreassen et al., 'Using probabilistic and decision-theoretic methods in treatment and prognosis modeling', *Artificial Intelligence in Medicine*, 1999; 15 (2): 121–34.
3. L. Leibovici, 'Alternative (complementary) medicine: a cuckoo in the nest of empiricist reed warblers', *British Medical Journal*, 1999; 319: 1629–32; Leibovici, 'Effects of remote, retroactive intercessory prayer', op. cit.
4. Letters, BMJ Online, December 22, 2003.
5. L. Dossey, 'How healing happens: exploring the nonlocal gap', *Alternative Therapies in Health and Medicine*, 2002; 8 (2): 12–16, 103–10.
6. B. Oshansky and L. Dossey, 'Retroactive prayer: A preposterous hypothesis?' *British Medical Journal*, 2003; 327: 20–7.
7. Letters, 'Effect of retroactive prayer', *British Medical Journal*, 2002; 324: 1037.
8. Correspondence from Liebovici to author, June 28, 2005.
9. Interview with Jahn and Dunne, July 2005.
10. R. G. Jahn et al., 'Correlations of random binary sequences with pre-stated operator intention: a review of a 12-year program', *Journal of Scientific Exploration*, 1997; 11 (3): 345–67.
11. D. J. Bierman and J. M. Houtkooper, 'Exploratory PK tests with a programmable high speed random number generator', *European Journal of Parapsychology*, 1975; 1 (1): 3–14.
12. R. Broughton, *Parapsychology: The Controversial Science*, New York: Ballantine Books, 1991: 175–6.
13. H. Schmidt and H. Stapp, 'Study of PK with prerecorded random events and the effects of preobservation', *Journal of Parapsychology*, 1993; 57: 351.
14. E. R. Gruber, 'Conformance behavior involving animal and human subjects', *European Journal of Parapsychology*, 1979; 3 (1): 36–50.
15. E. R. Gruber, 'PK effects on pre-recorded group behaviour of living systems', *European Journal of Parapsychology*, 1980; 3 (2): 167–75.

16. F. W. J. J. Snel and P. C. van der Sijde, 'The effect of retro-active distance healing on *Babeia rodhani* (rodent malaria) in rats', *European Journal of Parapsychology*, 1990; 8: 123–30.

17. W. Braud, unpublished study, 1993, as reported in W. Braud, 'Wellness implications of retroactive intentional influence: exploring an outrageous hypothesis', *Alternatives Therapies in Health and Medicine*, 2000; 6 (1): 37–48.

18. H. Schmidt, 'Random generators and living systems as targets in retro-PK experiments', *Journal of the American Society for Psychical Research*, 1997; 912 (1): 1–13.

19. D. Radin et al., 'Effects of distant healing intention through time and space: Two exploratory studies', *Proceedings of Presented Papers: The 41st Annual Convention of the Parapsychological Association*, Halifax, Nova Scotia, Canada: Parapsychological Association, 1998: 143–61.

20. J. R. Stroop, 'Studies of interference in serial verbal reactions', *Journal of Experimental Psychology*, 1935; 18: 643, as cited in D.I. Radin and E. C. May, 'Evidence for a retrocausal effect in the human nervous system', Boundary Institute Technical Report 2000–1.

21. H. Klintman, 'Is there a paranormal (precognitive) influence in certain types of perceptual sequences? Part I and II', *European Journal of Parapsychology*, 1983; 5: 19–49 and 1984; 5: 125–40, as cited in Radin and May, Boundary Institute Technical Report, op. cit.

22. Radin and May, Boundary Institute Technical Report, op. cit.

23. Braud, 'Wellness implications', op. cit.

24. See http://www.fourmilab.ch/rpkp/bierman-metaanalysis.html.

25. Radin and May, Boundary Institute Technical Report, op. cit.

26. G. A. Mourou and D. Umstadter, 'Extreme light', in 'The Edge of Physics'. Special edition of *Scientific American*, 2003; 13 (1): 77–83 updated from May 2002 issue.

27. L. H. Ford and T. A. Roman, 'Negative energy, wormholes and warp drive', in 'The Edge of Physics'. Special edition of *Scientific American*, 2003; 13 (1): 85–91 updated from January 2000 issue.

28. J. A. Wheeler and R. P. Reynman, 'Interaction with the absorber as the mechanism of radiation', *Reviews of Modern Physics*, 1945; 17 (2–3): 157–81; J. A. Wheeler and R. P. Reynman, 'Classical electrodynamics in terms of direct interparticle action', *Reviews of Modern Physics*, 1949; 21: 425–33.

29. E. H. Walker, 'The nature of consciousness', *Mathematical BioSciences*, 1970; 7: 131–78.

30. H. P. Stapp, 'Theoretical model of a purported empirical violation of the predictions of quantum theory', *Physical Review A*, 1994; 50 (1): 18–22.

31. Braud, 'Wellness implications', op. cit.

32. L. Grover, 'Quantum computing', *The Sciences*, July/August 1999: 24–30.

33. M. Brooks, 'The weirdest link', *New Scientist*, March 27, 2004; 181 (2440): 32–5.

34. D. Bierman, 'Do PSI-phenomena suggest radical dualism?' in S. Hammeroff et al. (ed.), *Toward a Science of Consciousness II*, Cambridge, Mass.: MIT Press, 1998: 709–14.

35. D. I. Radin, 'Experiments testing models of mind-matter interaction', *Journal of Scientific Exploration*, 2006; 20 (3), 375–401.

36. Interview with William Braud, October 1999.

37. W. Braud, 'Transcending the limits of time', *The Inner Edge: A Resource for Enlightened Business Practice*, 1999; 2 (6): 16–18.

38. R. D. Nelson, 'The physical basis of intentional healing systems', Technical Report, PEAR 99001, Princeton Engineering Anomalies Research, Princeton, New Jersey, January 1999.

39. Braud, interview with author, October 1999.

40. D. Bierman 'Does consciousness collapse the wave packet?' *Mind and Matter*, 2003; 1 (1): 45–58.

41. H Schmidt, 'Additional effect for PK on pre-recorded targets', *Journal of Parapsychology*, 1985; 49: 229–44; 'PK tests with and without preobservation by animals', in L. S. Henkel and J. Palmer (eds.), *Research in Parapsychology*, Metuchen, NJ: Scarecrow Press, 1990: 15–19.

Chapter 12: The Intention Experiment

1. Interview with Fritz-Albert Popp, March 1, 2006.

2. F.-A. Popp et al., 'Further analysis of delayed luminescence of plants', *Journal of Photochemistry and Photobiology B: Biology*, 2005, 78: 235–44.

3. For a full description of Popp's history, see McTaggart, *The Field*, op. cit.

4. International Institute of Biophysics, see www.lifescientists.de.

5. B. J. Dunne, 'Co-operator experiments with an REG device', PEAR Technical Note 91005, Princeton Engineering Anomalies Research, Princeton, New Jersey, December 1991.

6. R. D. Nelson et al., 'FieldREG anomalies in group situations', *Journal of Scientific Exploration*, 1996; 10 (1): 111–41; R. D. Nelson et al., 'FieldREGII: Consciousness field effects: replications and explorations', *Journal of Scientific Exploration*, 1998; 12 (3): 425–54.

7. D. I. Radin, 'For whom the bell tolls: A question of global consciousness', *Noetic Sciences Review*, 2003; 63: 8–13 and 44–5; R. D. Nelson et al., 'Correlation of continuous random data with major world events', *Foundations of Physics Letters*, 2002; 15 (6): 537–50.

8. D. I. Radin, 'Exploring relationships between random physical events and mass human attention: Asking for whom the bell tolls', *Journal of Scientific Exploration*, 2002; 16 (4): 533–47.

9. R. D. Nelson, 'Coherent consciousness and reduced randomness: Correlations on September 11, 2001', *Journal of Scientific Exploration*, 2002; 16 (4): 549–70.

10. Ibid.

11. Bryan J. Williams, 'Exploratory block analysis of field consciousness effects on global RNGs on September 11, 2001' (http://noosphere.princeton.edu/williams/GCP911.html).

12. J. D. Scargle, 'Commentary: Was there evidence of global consciousness on September 11, 2001?' *Journal of Scientific Exploration*, 2002; 16 (4): 571–7.

13. Nelson et al., 'Correlation of continuous random data', op. cit.

14. M. C. Dillbeck et al., 'The Transcendental Meditation program and crime rate change in a sample of 48 cities', *Journal of Crime and Justice*, 1981; 4: 25–45.

15. J. Hagelin et al., 'Effects of group practice of the Transcendental Meditation program on preventing violent crime in Washington, D.C.: Results of the National Demonstration Project, June–July 1993', *Social Indicators Research*, 1999; 47 (2): 153–201.

16. W. Orme-Johnson et al., 'International peace project in the Middle East: the effects of the Maharishi technology of the unified field', *Journal of Conflict Resolution*, 1988; 32: 776–812.

17. K. L. Cavanaugh et al., 'Consciousness and the quality of economic life; empirical research on the macroeconomic effects of the collective practice of Maharishi's Transcendental Meditation and TM-Sidhi program.' Paper originally presented at the annual meeting of the Midwest Management Society, Chicago, March 1989, published in R. G. Greenwood (ed.), *Proceedings of the Midwest Management Society*, Chicago: Midwest Management Society, 1989: 183–90; K. L. Cavanaugh et al., 'A multiple-input transfer function model of Okun's misery index: An empirical test of the Maharishi Effect.' Paper presented at the Annual Meeting of the American Statistical Association, Washington D.C., August 6–10, 1989, an abridged version of the paper appears in *Proceedings of the American Statistical Association, Business and Economics Statistics Section*, Alexandria, Va.: American Statistical Association, 1989: 565–70; K. L. Cavanaugh and K. D. King, 'Simultaneous transfer function analysis of Okun's misery index: improvements in the economic quality of life through Maharishi's Vedic Science and technology of consciousness.' Paper presented at the Annual Meeting of the American Statistical Association, New Orleans, Louisiana, August 22–25, 1988, an abridged version of the paper appears in *Proceedings of the American Statistical Association, Business and Economics Statistics Section*, Alexandria, Va.: American Statistical Association, 1988: 491–6; K. L. Cavanaugh, 'Time series analysis of U.S. and Canadian inflation and unemployment: A test of a field-theoretic hypothesis.' Paper presented at the Annual Meeting of the American Statistical Association, San Francisco, California, August 17–20, 1987, published in *Proceedings of the American Statistical Association, Business and Economics Statistics Section*, Alexandria, Va.: American Statistical Association, 1987: 799–804.

18. Strong rains fall on fire-ravaged Amazon state, March 31, 1998, Web posted at: 6:46 p.m. EST (2346 GMT), Brasilia, Brazil (CNN) http://twm.co.nz/.

19. R. Nelson, 'Wishing for good weather: a natural experiment in group consciousness', *Journal of Scientific Exploration*, 1997; 11 (1): 47–58.

20. M. Emoto, *The Hidden Messages in Water*, New York: Atria, 2005.

21. Interview with Dean Radin, May 3, 2006.

22. Not her real name. I've changed her name at her request. Nevertheless, our meditators were shown her real name and photo.

23. R. Van Wijk and E. P. Van Wijk, 'The search for a biosensor as a witness of a human laying on of hands ritual', *Alternative Therapies in Health and Medicine*, 2003; 9 (2): 48–55.

Chapter 13: The Intention Exercises

1. See C. T. Tart, 'Initial application of mindfulness extension exercises in a traditional Buddhist meditation retreat setting, 1995', unpublished (www.paradigm-sys.com/cttart).

2. R. McCraty et al., 'The electricity of touch: Detection and measurement of cardiac energy exchange between people', in K. H. Pribram (ed.), *Brain and Values: Is a Biological Science of Values Possible?* Mahwah, NJ: Lawrence Erlbaum Associates, 1998: 359–79.

3. S. Rinpoche, *The Tibetan Book of Living and Dying*, San Francisco: HarperSanFrancisco, 1994.
4. S. Rinpoche, as quoted in J. Stone, Instructor's Training Manual, Course Syllabus: Training in Compassionate-Loving Intention, 2003.
5. H. Dienstfrey, *Where the Mind Meets the Body*, London: HarperCollins, 1991: 39.

Afterword: Thoughts Heard 'Round the World: The First Intention Experiments

1. Dr. Schwartz has recounted all his work in his book *The Energy Healing Experiments: Science Reveals Our Natural Power to Heal* (Atria, 2007).
2. B. Grad, 'A telekinetic effect on plant growth,' International Journal of Parapsychology, 11963; 5: 117–33; B. Grad,' A telekinetic effect on plant growth: II. Experiments involving treatment of saline in stopped bottles,' International Journal of Parapsychology, 1964; 6: 473–98.
3. S. M. Roney-Dougal and J. Solfvin, 'Filed study of enhancement effect on lettuce seeds – their germination rate, growth and health,' Journal of the Society for Psychical Research, 2002; 66: 129–43.
4. S. M. Roney-Dougal, 'Field study of an enhancement effect on lettuce seeds: a replication study,' Journal of Parapsychology, 2003:67: 279–98.

Bibliography

Editorial, 'MANTRA II: Measuring the unmeasurable?' *The Lancet*, 2005; 366 (9481): 178.

'New spin on salt', *University of Chicago Magazine*, August 2004, 96 (6), http://magazine.uchicago.edu/0408/research/spin.shtml.

'Science Fact: Scientists achieve "Star Trek"-like feat', The Associated Press, December 10, 1997, posted on CNN, http://edition.cnn.com/TECH/9712/10/beam.me.up.ap.

'Strong rains fall on fire-ravaged Amazon state, March 31, 1998', Web posted at: 6:46 p.m. EST (23:46 GMT), Brasilia, Brazil (CNN) http://twm.co.nz/.

Achterberg, J. and Lawlis, G. F., *Bridges of the Bodymind: Behavioral Approaches for Health Care*, Champaign, Ill.: Institute for Personality and Ability Testing, 1980.

Achterberg, J. et al., 'Evidence for correlations between distant intentionality and brain function in recipients: a functional magnetic resonance imagining analysis', *Journal of Alternative and Complementary Medicine*, 2005; 11 (6): 965–71.

Adams, M. H., 'Variability in remote-viewing performance: possible relationship to the geomagnetic field', in D. H. Weiner and D. I. Radin (eds.), *Research in Parapsychology*, Metuchen, NJ: Scarecrow Press, 1986.

Alexander, C. et al., 'EEG and SPECT data of a selected subject during psi tests: the discovery of a neurophysiological correlate', *Journal of Parapsychology*, 1998; 62 (20): 102–4.

Allahverdiyev, A. R. et al., 'Possible space weather influence on functional activity of the human brain.' Paper presented at Space Weather Workshop: Looking Toward a European Space Weather Programme, November 17–19, 2001, ESTEC, Noordwijk, the Netherlands.

Andreassen, S. et al., 'Using probabilistic and decision-theoretic methods in treatment and prognosis modeling', *Artificial Intelligence in Medicine*, 1999; 15 (2): 121–34.

Arndt, M. et al., 'Probing the limits of the quantum world', *Physics World*, March 2005.

Arndt, M. et al., 'Wave–particle duality of C60 molecules', *Nature*, 1999; 401: 680–2.

Arnesen, C. et al., 'Thermal and magnetic entanglement in the 1D Heisenberg Model', *Physical Review Letters*, 2001; 87: 017901.

Astin, J. et al., 'The efficacy of "distant healing": a systematic review of randomized trials', *Annals of Internal Medicine*, 2000; 132: 903–10.

Astin, J. A., 'Mind-body therapies for the management of pain', *Clinical Journal of Pain*, 2004; 20 (1): 27–32.

Astin, J. A. et al., 'Mind-body medicine: state of the science: implications for practice', *Journal of the American Board of Family Practitioners*, 2003; 16 (2): 131–47.

Atmanspacher, H., 'Mind and matter as asymptotically disjoint, inequivalent representations with broken time-reversal symmetry', *BioSystems*, 2003; 68: 19–30.

Auerbach, L., *Mind Over Matter: A Comprehensive Guide to Discovering Your Psychic Powers*, New York: Kensington Books, 1996.

Aviles, J. M. et al., 'Intercessory prayer and cardiovascular disease progression in a coronary care unit population: a randomized controlled trial', *Mayo Clinic Proceedings*, 2001; 76 (12): 1192–8.

Babayev, E., 'Some results of investigations on the space weather influence on functioning of several engineering-technical and communication systems and human health', *Astronomical and Astrophysical Transactions*, 2003; 22 (6): 861–7.

Backster, C., 'Evidence of a primary perception in plant life', *International Journal of Parapsychology*, 1968; 10 (4): 329–48.

Backster, C., *Primary Perception: Biocommunication with Plants, Living Foods, and Human Cells*, Anza, Calif.: White Rose Millennium Press, 2003.

Ban, M., 'Measurement-induced enhancement of entanglement of a two-mode squeezed-vacuum state.' Letter to the editor, *Journal of Optics B: Quantum and Semiclassical Optics*, 2005; 7: L4–L7.

Baraz, J. and Tart, C. T., 'Initial application of mindfulness extension exercises in a traditional Buddhist meditation retreat setting', unpublished, 1995 © C. Tart and J. Baraz.

Barber, T. X., 'Changing "unchangeable" bodily processes by (hypnotic) suggestions: a new look at hypnosis, cognitions, imagining and the mind-body problem', in A. A. Sheikh (ed.), *Imagination and Healing*, Farmingdale, NY: Baywood Publishing Co., 1984.

Baroga, L., 'Influence on the sporting result of the concentration of attention process and time taken in the case of weight lifters', in *Proceedings of the Third World Congress of the International Society of Sports Psychology*, Volume 3, Madrid, Spain: Instituto Nacional de Educacion Fisica Y Deportes, 1973.

Barr, K. and Hall, C., 'The use of imagery by rowers', *International Journal of Sport Psychology*, 1992; 23: 243–61.

Barrett, J., 'Going the distance', *Intuition*, June/July 1999: 30–1.

Basar-Eroglu, C., 'Gamma-band responses in the brain: a short review of psychophysiological correlates and functional significance', *International Journal of Psychophysiology*, 1996; 24 (1–2): 101–2.

Bell, I. R. et al., 'Development and validation of a new global well-being outcomes rating scale for integrative medicine research', *BMC Complementary and Alternative Medicine*, 2004; 4: 1.

Bell, I. R. et al., 'Gas discharge visualization evaluation of ultramolecular doses of homeopathic medicines under blinded, controlled conditions', *Journal of Alternative and Complementary Medicine*, 2003; 9 (1): 25–38.

Beloussov, L. and Louchinskaia, N. N., 'Biophoton emission from developing eggs and embryos: nonlinearity, wholistic properties and indications of energy transfer', in J. J. Chang et al. (eds.), *Biophotons*, London: Kluwer Academic Publishers, 1998: 121–40.

Benor, D. J., *Spiritual Healing: Scientific Validation of a Healing Revolution*, Southfield, Mich.: Vision Publications, 2001.

Benor, D. J., *Spiritual Healing: Scientific Validation of a Healing Revolution Professional Supplement*, Southfield, Mich.: Vision Publications, 2002.

Benor, D. J., *Healing Research: Holistic Energy Medicine and Spirituality*, 4 vols., Deddington, Oxfordshire: Helix Editions Ltd, 1993.

Benson H., 'Body temperature changes during the practice of g Tum-mo yoga (Matters Arising)', *Nature*, 1982; 298: 402.

Benson, H. et al., 'Body temperature changes during the practice of g tum-mo (heat) yoga', *Nature*, 1982; 295: 234–6.

Benson, H. et al., 'Decreased systolic blood pressure through operant conditioning techniques in patients with essential hypertension', *Science*, 1971; 173 (3998): 740–2.

Benson, H. et al., 'Study of the therapeutic effects of intercessory prayer (STEP) in cardiac bypass patients: a multi-center randomized trial of uncertainty and certainty of receiving intercessory prayer', *American Heart Journal*, 2006; 151 (4): 934–42.

Benson, H. et al., 'Three case reports of the metabolic and electroencephalographic changes during advanced Buddhist meditation techniques', *Behavioral Medicine*, 1990; 16 (2): 90–5.

Bernardi, L. et al., 'Effect of rosary prayer and yoga mantras on autonomic cardiovascular rhythms: comparative study', *British Medical Journal*, 2001; 323: 1446–9.

Bierman, D., 'Do PSI-phenomena suggest radical dualism?' in S. Hammeroff et al. (eds.), *Toward a Science of Consciousness II*, Cambridge, Mass.: The MIT Press, 1998: 709–14.

Bierman, D. J. and Houtkooper, J. M., 'Exploratory PK tests with a programmable high speed random number generator', *European Journal of Parapsychology*, 1975; 1–1: 3–14.

Binhi, V. N. and Savin, A. V., 'Molecular gyroscopes and biological effects of weak extremely low-frequency magnetic fields', *Physical Review E*, 2002; 65: 051912–22.

Blasband, R., 'The ordering of random events by emotional expression', *Journal of Scientific Exploration*, 2000; 14 (2): 195–216.

Blasband, R. A., 'Working with the body in psychotherapy from a Reichian viewpoint', *AHP Perspective*, June 2005.

Blasband, R. A. and Martin, G., 'Biophoton emission in "orgone energy" treated cress seeds, seedlings and *Acetabularia*', International Consciousness Research Laboratory, ICRL Report No 93.6.

Booth, J. N. et al., 'Ranking of stimuli that evoked memories in significant others after exposure to circumcerebral magnetic fields: correlations with ambient geomagnetic activity', *Perceptual and Motor Skills*, 2002; 95 (2): 555–8.

Bose, S., 'Multiparticle generation of entanglement swapping', *Physical Review A*, 1998; 57 (2): 822–9.

Bratman, M. E., 'What is intention?' in M. Pollack, P. Cohen and J. L. Morgan (eds.), *Intentions in Communication*, Cambridge, Mass.: The MIT Press, 1990: 15–31.

Braud, W., 'Transcending the limits of time', *The Inner Edge: A Resource for Enlightened Business Practice*, 1999; 2 (6): 16–18.

Braud, W., 'Wellness implications of retroactive intentional influence: exploring an outrageous hypothesis', *Alternative Therapies in Health and Medicine*, 2000; 6 (1): 37–48.

Braud, W. and Schlitz, M., 'Psychokinetic influence on electrodermal activity', *Journal of Parapsychology*, 1983; 47 (2): 95–119.

Braud, W. et al., 'Experiments with Matthew Manning', *Journal of the Society for Psychical Research*, 1979; 50: 199–223.

Braud, W. G. 'Can our intentions interact directly with the physical world?' *European Journal of Parapsychology*, 1994; 10: 78–90.

Braud, W. G. and Dennis, S. P., 'Geophysical Variables and Behavior, LVIII: Autonomic activity, hemolysis and biological psychokinesis: possible relationships with geomagnetic field activity', *Perceptual and Motor Skills*, 1989; 68: 1243–54.

Braud, W. G. and Schlitz, M. J., 'A method for the objective study of transpersonal imagery', *Journal of Scientific Exploration*, 1989; 3 (1): 43–63.

Braud W. G. and Schlitz, M. J., 'Consciousness interactions with remote biological systems: anomalous intentionality effects', *Subtle Energies and Energy Medicine*, 1991; 2 (1): 1–27.

Braud, W. G. et al., 'Further studies of autonomic detection of remote staring: replication, new control procedures and personality correlates', *Journal of Parapsychology*, 1993; 57: 391–409.

Braud, W. G. et al., 'Further studies of the bio-PK effect: feedback, blocking specificity/generality', in R. White and J. Solfvin (eds.), *Research in Parapsychology*, Metuchen, NJ: Scarecrow Press, 1984: 45–8.

Brooks, M., 'Curiouser and curiouser', *New Scientist*, 2003; 178 (2394): 28.

Brooks, M., 'Entanglement: the weirdest link', *New Scientist*, 2004; 181 (2440): 32–5.

Broughton, R. S., *Parapsychology: The Controversial Science*, New York: Ballantine Books, 1991.

Brown, D. et al., 'Differences in visual sensitivity among mindfulness meditators and non-meditators', *Perceptual and Motor Skills*, 1984; 58 (3): 775–84.

Brukner, C., 'Quantum entanglement in time', http://arxiv.org/abs/quant-ph/0402127.

Brukner, C. et al., 'Crucial role of quantum entanglement in bulk properties of solids', *Physical Review A*, 2006; 73: 012100–4.

Buccheri, R. et al. (eds.), *Abstracts of Talks*, 'Endophysics, Time, Quantum and the Subjective.' ZiF interdisciplinary research workshop, January 17–22, 2005, Bielefeld, Germany.

Bundzen, P. V. et al., 'Altered states of consciousness; review of experimental data obtained with a multiple techniques approach', *Journal of Alternative and Complementary Medicine*, 2002; 8 (2): 153–65.

Bundzen, P. V. et al., 'Psychophysiological correlates of athletic success in athletes training for the Olympics', *Human Physiology*, 2005; 31 (3): 316–23.

Bunnell, T., 'A tentative mechanism for healing', *Positive Health*, December 1997; 23.

Bunnell, T., 'The effect of "healing with intent" on pepsin enzyme activity', *Journal of Scientific Exploration*, 1999; 13 (2): 139–48.

Bureau, Y. and Persinger, M., 'Decreased latencies for limbic seizures induced in rats by lithium-pilocarpine occur when daily average geomagnetic activity exceeds 20 nanotesla', *Neuroscience Letters*, 1995; 192: 142–4.

Bureau, Y. and Persinger, M., 'Geomagnetic activity and enhanced mortality in rats with acute (epileptic) limbic lability', *International Journal of Biometeorology*, 1992; 36: 226–32.

Burhans, R. S. et al., 'Mental imagery training: effects on running speed performance', *International Journal of Sport Psychology*, 1988; 19: 26–37.

Burleson, K. O. et al., 'Energy healing training and heart rate variability.' Letter to editor, *Journal of Alternative and Complementary Medicine*, 2005; 11 (3): 391–5.

Byrd, R. C., 'Positive therapeutic effects of intercessory prayer in a coronary care unit population', *Southern Medical Journal*, 1988, 81 (7): 826–9.

Cautela, J. R. and Kearney, A. J. (eds.), *Covert Conditioning Casebook*, Boston, Mass.: Thomson Brooks/Cole, 1993.

Cavanaugh, K. L., 'Time series analysis of U. S. and Canadian inflation and unemployment: a test of a field-theoretic hypothesis', in *Proceedings of the American Statistical Association, Business and Economics Statistics Section*, Alexandria, Va.: American Statistical Association, 1987: 799–804.

Cavanaugh, K. L. and King, K. D., 'Simultaneous transfer function analysis of Okun's misery index: improvements in the economic quality of life through Maharishi's Vedic science and technology of consciousness', in *Proceedings of the American Statistical Association, Business and Economics Statistics Section*, Alexandria, Va.: American Statistical Association, 1988: 491–6.

Cavanaugh, K. L. et al., 'A multiple-input transfer function model of Okun's misery index: an empirical test of the Maharishi effect', in *Proceedings of the American Statistical Association, Business and Economics Statistics Section*, Alexandria, Va.: American Statistical Association, 1989: 565–70.

Cavanaugh, K. L. et al., 'Consciousness and the quality of economic life; empirical research on the macroeconomic effects of the collective practice of Maharishi's Transcendental Meditation and TM-Sidhi program', in R. G. Greenwood (ed.), *Proceedings of the Midwest Management Society*, Chicago, Ill.: Midwest Management Society, 1989: 183–90.

Chang, J. J. et al., 'Communication between dinoflagellates by means of photon emission', in L. V. Beloussov and F.-A. Popp (eds.), *Proceedings of International Conference on Non-equilibrium and Coherent Systems in Biophysics, Biology and Biotechnology, Sep. 28–Oct.2, 1994*, Moscow: Bioinform Services Co., 1995: 318–30.

Chang, J. J. et al., 'Research on cell communication of *P. elegans* by means of photon emission', *Chinese Science Bulletin*, 1995; 40: 76–9.

Chase, M. H. et al., 'Afferent vagal stimulation: neurographic correlates of induced EEG synchronization and desynchronization', *Brain Research*, 1967; 5: 236–49.

Chen, Z. B. et al., 'All-versus-nothing violation of local realism for two entangled photons', *Physical Review Letters*, 2003; 90: 160408.

Claghorn, J. L., 'Directional effects of skin temperature self-regulation on regional cerebral blood flow in normal subjects and migraine patients', *American Journal of Psychiatry*, 1981; 138 (9): 1182–7.

Clemente, C. D. et al., 'Postreinforcement EEG synchronization during alimentary behavior', *Electroencephalography and Clinical Neurophysiology*, 1964; 16: 335–65.

Co. S. and Robins, E. B., *Your Hands Can Heal You*, New York: Free Press, 2002.

Cohen, K. S., *The Way of Qigong: The Art and Science of Chinese Energy Healing*, New York: Bantam, 1997.

Cohen, P., 'Mental gymnastics', *New Scientist*, November 24, 2001; 172 (2318): 17.

Cohen S. and Popp, F.-A., 'Biophoton emission of the human body', *Journal of Photochemistry and Photobiology*, 1997; 40: 187–9.

Cohen, S. et al., 'Non-local effects of biophoton emission from the human body', www.lifescientists.de.

Connor, M., 'Baseline testing of energy practitioners: biophoton imaging results.' Paper presented at the North American Research in Integrative Medicine conference, Edmonton, Canada, May 2006.

Connor, M. et al., 'Oscillation of amplitude as measured by an extra low frequency magnetic field meter as a biophysical measure of intentionality.' Paper presented at the Toward a Science of Consciousness Conference, Tuscon, Arizona, April 2006.

Cooperstein, M. A., 'The myths of healing: a summary of research into transpersonal healing experience', *Journal of the American Society for Psychical Research*, 1992; 86: 99–133.

Corby, J. C. et al., 'Psychophysiological correlates of the practice of Tantric Yoga meditation', *Postgraduate Medical Journal*, 1985; 61: 301–4.

Cornélissen, G. et al., 'Chronomes, time structures, for chronobioengineering for "a full life"', *Biomedical Instrumentation and Technology*, 1999; 33 (2): 152–87.

Cornélissen, G. et al., 'Is a birth-month-dependence of human longevity influenced by half-yearly changes in geomagnetics?' 'Physics of Auroral Phenomena', Proceedings, XXV Annual Seminar, Apatity: Polar Geophysical Institute, Kola Science Center, Russian Academy of Science, February 26–March 1; 2002; 161–6.

Cornélissen, G. et al., 'Non-photic solar associations of heart rate variability and myocardial infarction', *Journal of Atmospheric and Solar-terrestrial Physics*, 2002; 64: 707–20. Creath, K. and Schwartz, G. E., 'What biophoton images of plants can tell us about biofields and healing', *Journal of Scientific Exploration*, 2005; 19 (4): 531–50.

Creath, K., 'Biophoton images of plants: revealing the light within', *Journal of Alternative and Complementary Medicine*, 2004; 10 (1): 23–6.

Creath, K. and Schwartz, G. E., 'Measuring effects of music, noise, and healing energy using a seed germination bioassay', *Journal of Alternative and Complementary Medicine*, 2004; 10 (1): 113–22.

Crombie, W. J., 'Meditation changes temperatures: mind controls body in extreme experiments', *Harvard University Gazette*, April 18, 2002.

Damasio, A. R., *Descartes' Error*, New York: Grosset-Putnam, 1994.

Das N. and Gastaut H., 'Variations in the electrical activity of the brain, heart and skeletal muscles during yogic meditation and trance', *Electroencephalography and Clinical Neurophysiology*, 1955, Supplement no 6: 211–19.

Davidson, R. J., 'Alterations in brain and immune function produced by mindfulness meditation', *Psychosomatic Medicine*, 2003; 65: 564–70.

Davidson, R. J. and van Reekum, C. M., 'Emotion is not one thing', *Psychological Inquiry*, 2005; 16: 16–18.

Davidson, R. J. et al., 'Alterations in brain and immune function produce by mindfulness meditation', *Psychosomatic Medicine*, 2003; 65: 564–70.

Davis, K. et al., 'Bose-Einstein condensation in a gas of sodium atoms', *Physical Review Letters*, 1995; 75: 3969–73.

de la Fuente-Fernández, R. et al., 'Expectation and dopamine release: mechanism of the placebo effect in Parkinson's disease', *Science*, 2001: 293 (5532): 1164–6.

Delanoy, D. et al., 'An EDADMILS study exploring agent-receiver pairing', *Proceedings of Presented Papers*, The Parapsychological Association, 42nd Annual Convention, 1999: 68–82.

Denis, M., 'Visual imagery and the use of mental practice in the development of motor skills', *Canadian Journal of Applied Sport Sciences*, 1985; 10: 4S–16S.

Dennett, D., 'Three kinds of intentional psychology', in *The Intentional Stance*, Cambridge, Mass.: The MIT Press, 1987: 43–68.

Derr, J. S. and Persinger, M. A., 'Geophysical Variables and Behavior, LIV: Zeitoun (Egypt) apparitions of the Virgin Mary as tectonic strain-induced luminosities', *Perceptual and Motor Skills*, 1989; 68: 123–8.

Diamond, J., *Life Energy*, New South Wales: Angus & Robertson, 1992.

Diamond, J., *Your Body Doesn't Lie*, New York: HarperCollins, 1979.

Dibble, W. E. and Tiller, W. A., 'Electronic device-mediated pH changes in water', *Journal of Scientific Exploration*, 1999; 13: 2–10.

Dienstfrey, H., *Where the Mind Meets the Body*, London: HarperCollins, 1991.

Dienstfrey, H., 'Mind and mindlessness in mind–body research', in M. Schlitz et al. (eds.), *Consciousness and Healing: Integral Approaches to Mind–Body Healing*, St Louis, Mo.: Elsevier Churchill Livingstone, 2005: 51–60.

Dillbeck, M. C. et al., 'The Transcendental Meditation program and crime rate change in a sample of 48 cities', *Journal of Crime and Justice*, 1981; 4: 25–45.

Dobie, T. G., 'A comparison of two methods of training resistance to visually-induced motion sickness.' Paper presented at VII International Man in Space Symposium: Physiologic Adaptation of Man in Space, Houston, Texas, 1986. *Aviation, Space, and Environmental Medicine*, 1987; 58 (9), Sect. 2: 34–41.

Dossey, L., *Be Careful What You Pray For … You Just Might Get It*, San Francisco: HarperSanFrancisco, 1997.

Dossey, L., 'Commentary', *Archives of Internal Medicine*, 2000; 160; 1735–8.

Dossey, L., *Healing Words: The Power of Prayer and the Practice of Medicine*, San Francisco: HarperSanFrancisco, 1993.

Dossey, L., 'How healing happens: exploring the nonlocal gap', *Alternative Therapies in Health and Medicine*, 2002; 8 (2): 12–16, 103–110.

Dossey, L., *Meaning and Medicine: Lessons from a Doctor's Tales of Breakthrough Healing*, New York: Bantam, 1991.

Dossey, L., 'Prayer experiments: science or folly? Observations on the Harvard prayer study', *Network Review* (UK), 2006; 91: 22–3.

Duane, T. D. and Behrendt, T., 'Extrasensory electroencephalographic induction between identical twins', *Science*, 1965; 150: 367.

Dubrov, A. P., 'Distant mental healing: influence of intercessory prayers and qi-gong therapy', *The International Journal of Healing and Caring On-line*, 2005; 5 (3).

Dubrov, A. P. and Pushkin, V. N., *Parapsychology and Contemporary Science*, New York and London: Consultants Bureau, 1982.

Dunn, B. R. et al., 'Concentration and mindfulness meditations: unique forms of consciousness?' *Applied Psychophysioloical Biofeedback*, 1999; 24 (3): 147–65.

Dunne, B. J., 'Co-operator experiments with an REG device', PEAR Technical Note 91005, Princeton Engineering Anomalies Research, Princeton, New Jersey, December 1991.

Early, L. F. and Kifschutz, J. E., 'A case of stigmata', *Archives of General Psychiatry*, 1974; 30: 197–200.

Ebisch, R., 'It's all in the timing', *Sky Magazine*, 1995.

Edelman, G. M. and Tononi, G., *Consciousness: How Matter Becomes Imagination*, London: Penguin, 2000.

Eden, D., *Energy Medicine*, London: Piatkus, 1998.

Ekman, P. et al., 'Buddhist and psychological perspectives on emotions and well-being', *Current Directions in Psychological Science*, 2005; 14: 59–63.

Eller, L. S., 'Guided imagery interventions for symptom management', *Annual Review of Nursing Research*, 1999; 17: 57–84.

Emoto, M., *The Hidden Messages in Water*, New York: Atria, 2005.

Fahrion, S. et al., 'Biobehavioral treatment of essential hypertension: a group outcome study', *Biofeedback and Self Regulation*, 1986; 11 (4): 257–77.

Fahrion, S. et al., 'EEG amplitude, brain mapping and synchrony in and between a bioenergy practitioner and client during healing', *Subtle Energies and Energy Medicine*, 1992; 3 (1): 19–52.

Feltz, D. L. et al., 'A revised meta-analysis of the mental practice literature on motor skill learning', in D. Druckman and J. A. Swets (eds.), *Enhancing Human Performance: Issues, Theories, and Techniques*, Washington, DC: National Academy Press, 1988.

Fenwick, P. B., 'Metabolic and EEG changes during transcendental meditation: an explanation', *Biological Psychology*, 1977; 5 (2): 101–18.

Feynman, R. P., *Six Easy Pieces: The Fundamentals of Physics Explained*, London: Penguin, 1995.

Ford, L. H. and Roman, T. A., 'Negative energy, wormholes and warp drive', special edition of *Scientific American*, 2003; 13 (1): 77–83.

Francomano, C. A., Jonas, W. B. and Chez, R. A. (eds.), *Proceedings: Measuring the Human Energy Field: State of the Science*. The Gerontology Research Center, National Institute of Aging, National Institutes of Health, Baltimore, Maryland, April 17–18, 2002.

Frantzis, B. K., *Opening the Energy Gates of Your Body*, Berkeley, Calif.: Blue Snake Books, 2006.

Frantzis, B. K., *The Water Method of Taoist Meditation, Volume 1: Relaxing Into Your Being: Breathing, Chi and Dissolving the Ego*, Berkeley, Calif.: North Atlantic Books, 1998.

Frantzis, B. K., *The Water Method of Taoist Meditation, Volume 2: The Great Stillness: Body Awareness, Moving Meditation and Sexual Chi Gung*, Berkeley, Calif.: North Atlantic Books, 1999.

Freedman, M. et al., 'Effects of frontal lobe lesions on intentionality and random physical phenomena', *Journal of Scientific Exploration*, 2003; 17 (4): 651–68.

Freeman, W. J., *How Brains Make Up their Minds*, London: Orion Books, 1999.

Friedman, H. et al., 'Geomagnetic parameters and psychiatric hospital admissions', *Nature*, 1963; 200: 626–8.

Fröhlich, H., 'Long range coherence and energy storage in biological systems', *Int. J. Quantum Chem.*, 1968; II: 641–9.

Fujita, A., 'An experimental study on the theoretical basis of mental training', in *Proceedings of the 3rd World Congress of the International Society of Sports Psychology: Volume Abstracts*, Madrid, Spain: Instituto Nacional de Educacion Fisica Y Deportes, 1973: 37–8.

Galle, R. M. et al., 'Biophoton emission from *Daphnia magna*: a possible factor in the self-regulation of swarming', *Experientia*, 1991; 47: 457–460.

Gershon, M., *The Second Brain: A Groundbreaking New Understanding of Nervous Disorders of the Stomach and Intestine*, London: HarperCollins, 1999.

Ghosh, S. et al., 'Coherent spin oscillations in a disordered magnet', *Science*, 2002; 296: 2195–8.

Ghosh, S. et al., 'Entangled quantum state of magnetic dipoles', *Nature*, 2003; 435: 48–51.

Gissurarson, L. R., 'The psychokinesis effect: geomagnetic influence, age and sex differences', *Journal of Scientific Exploration*, 1992; 6 (2): 157–65.

Gnevyshev, M. N., 'Essential features of the 11-year solar cycle', *Solar Physics*, 1977; 51: 175–82.

Goleman, D., *Destructive Emotions and How We Can Overcome Them*, London: Bloomsbury Press, 2004.

Goleman, D., *Emotional Intelligence*, London: Bloomsbury Press, 1996.

Goleman, D., 'Meditation and consciousness: an Asian approach to mental health', *American Journal of Psychotherapy*, 1976; 30 (1): 41–54.

Goleman, D., 'Why the brain blocks daytime dreams', *Psychology Today*, 1976; March: 69–71.

Grad, B., 'Dimensions in "Some biological effects of the laying on of hands" and their implications', in H. A. Otto and J. W. Knight (eds.), *Dimension in Wholistic Healing: New Frontiers in the Treatment of the Whole Person*, Chicago, Ill.: Nelson-Hall, 1979: 199–212.

Grad, B., 'Science investigates laying on of hands', Proceedings of 'Mind in Search of Itself', Mind Science Foundation and Silva International, Washington, DC, November 25–26, 1972.

Grad, B., 'The "laying on of hands": implications for psychotherapy, gentling and the placebo effect', *Journal of the Society for Psychical Research*, 1967; 61 (4): 286–305.

Green, E. E., 'Copper wall research psychology and psychophysics: subtle energies and energy medicine: emerging theory and practice', Proceedings, First Annual Conference, International Society for the Study of Subtle Energies and Energy Medicine (ISSSEEM), Boulder, Colorado, 21–25 June, 1991.

Green, E. E., 'Feedback technique for deep relaxation', *Psychophysiology*, 1969; 6 (3): 371–7.

Green, E. E. et al., 'Anomalous electrostatic phenomena in exceptional subjects', *Subtle Energies and Energy Medicine*, 1993; 2: 69.

Green, E. E. et al., 'Self-regulation of internal states', in J. Rose (ed.), *Progress of Cybernetics: Proceedings of the First International Congress of Cybernetics, London, September 1969*, London: Gordon and Breach Science Publishers, 1970: 1299–1318.

Green, E. E. et al., 'Voluntary control of internal states: psychological and physiological', *Journal of Transpersonal Psychology*, 1970; 2: 1–26.

Greyson, B., 'Distance healing of patients with major depression', *Journal of Scientific Exploration*, 1996; 10 (4): 447–65.

Gribbin, J., *Q is for Quantum: Particle Physics from A to Z*, London: Phoenix Giant, 1999.

Grinberg-Zylberbaum, J. and Ramos, J., 'Patterns of interhemisphere correlations during human communication', *International Journal of Neuroscience*, 1987; 36: 41–53.

Grinberg-Zylberbaum, J. et al., 'Human communication and the electrophysiological activity of the brain', *Subtle Energies and Energy Medicine*, 1992; 3 (3): 25–43.

Grinberg-Zylberbaum, J. et al., 'The Einstein-Podolsky-Rosen paradox in the brain: the transferred potential', *Physics Essays*, 1994; 7 (4): 422–8.

Grover, L., 'Quantum computing', *The Sciences*, July/August 1999: 24–30.

Gruber, E. R., 'Conformance behavior involving animal and human subjects', *European Journal of Parapsychology*, 1979; 3 (1): 36–50.

Gruber, E. R., 'PK effects on pre-recorded group behaviour of living systems', *European Journal of Parapsychology*, 1980; 3 (2): 167–75.

Gunlycke, D., 'Thermal concurrence mixing in a one-dimensional Ising model', *Physical Review A*, 2001; 64: 042302–9.

Gurfinkel, I. et al., 'Assessment of the effect of a geomagnetic storm on the frequency of appearance of acute cardiovascular pathology', *Biofizika*, 1998; 43 (4): 654–8.

Hackermueller, L., 'The wave nature of biomolecules and fluorofullerenes', *Physical Review Letters*, 2003; 91: 090408.

Hadhazy, V. A. et al., 'Mind-body therapies for the treatment of fibromyalgia. A systematic review', *Journal of Rheumatology*, 2000; 27 (12): 2911–18.

Hagan, S. et al., 'Quantum computation in brain microtubules: decoherence and biological feasibility', *Physical Review E*, 2002; 65: 061901–11.

Hagelin, J. et al., 'Effects of group practice of the Transcendental Meditation program on preventing violent crime in Washington, D. C.: results of the National Demonstration Project, June–July 1993', *Social Indicators Research*, 1999; 47 (2): 153–201.

Hagen, S., *Buddhism Plain and Simple*, New York: Broadway Books, 1999.

Haisch, B., Rueda, A. and Puthoff, H. E., 'Inertia as a zero-point-field Lorentz force', *Physical Review A*, 1994; 49 (2): 678–94.

Haisch, B., Rueda, A. and Puthoff, H. E., 'Physics of the zero-point field: implications for inertia, gravitation and mass', *Speculations in Science and Technology*, 1997; 20: 99–114.

Halberg, F., 'Transdisciplinary unifying implications of circadian findings in the 1950s', *Journal of Circadian Rhythms*, 2003; 1: 2.

Halberg, F. et al., 'Cross-spectrally coherent about 10-5- and 21-year biological and physical cycles, magnetic storms and myocardial infarctions', *Neuroendrocrinology Letters*, 2000; 21: 233–58.

Hall, S. S., 'Is Buddhism good for your health?' *New York Times Magazine*, September 14, 2003: 47–9.

Hameroff, S. R., 'Cytoplasmic gel states and ordered water: possible roles in biological quantum coherence.' Proceedings of the 2nd Annual Advanced Water Sciences Symposium, Dallas, Texas, 1996.

Hameroff, S. R. et al. (eds.), *Toward a Science of Consciousness II: The Second Tucson Discussions and Debate*, Cambridge, Mass.: The MIT Press, 1998.

Harrington, A. (ed.), *The Placebo Effect: An Interdisciplinary Exploration*, Cambridge, Mass.: Harvard University Press, 1997.

Harris, W. et al., 'A randomised, controlled trial of the effects of remote, intercessory prayer on outcomes in patients admitted to the coronary care unit', *Archives of Internal Medicine*, 1999; 159 (19): 2273–8.

Henderson, M., 'Hypnosis really does turn black into white', *The Times*, February 18, 2002.

Hercz, R., 'The God helmet', *SATURDAYNIGHT* magazine, October 2002: 40–6.

Hillman, D. et al., 'About-10 yearly (Circadecennian) cosmo-helio geomagnetic signatures in *Acetabularia*', *Scripta Medica* (BRNO), 2002; 75 (6): 303–8.

Hinshaw, K. E., 'The effects of mental practice on motor skill performance: critical evaluation and meta-analysis', *Imagination, Cognition and Personality*, 1991–2; 11: 3–35.

Hitt, J., 'This is your brain on God', *Wired*, November 1999; issue 7.11.

Hodges, R. D. and Schofield, A. M., 'Is spiritual healing a valid and effective therapy?' *Journal of the Royal Society of Medicine*, 1995; 88: 2033–7.

Holmes, E., *Living the Science of Mind*, Marina del Rey, Calif.: DeVorss & Company, 1984.

Holmes, R., 'In search of God', *New Scientist*, April 21, 2001; 2287.

Ikemi, A. et al., 'Thermographical analysis of the warmth of the hands during the practice of self-regulation method', *Psychotherapy and Psychosomatics*, 1988; 50 (1): 22–8.

Jahn, R. G. et al., 'Correlations of random binary sequences with pre-stated operator intention: a review of a 12-year program', *Journal of Scientific Exploration*, 1997; 11 (3): 345–67.

January Bishop, J. P. and Stenger, V. J., 'Retroactive prayer: lots of history, not much mystery, and no science', *British Medical Journal*, 2004; 329: 1444–6.

Jibu, M. and Yasue, K., *Quantum Brain Dynamics and Consciousness*, Amsterdam and Philadelphia: John Benjamins Publishing Company, 1995.

Jibu, M. et al., 'Quantum optical coherence in cytoskeletal microtubules: implications for brain function', *Biosystems*, 1994; 32: 195–209.

Josephson, B. D. and Pallikari-Viras, F., 'Biological utilisation of quantum nonlocality', *Foundations of Physics*, 2001; 21: 197–207.

Kamiya, J., 'Operant control of the EEG alpha rhythm', in C. Tart (ed.), *Altered States of Consciousness*, New York: John Wiley & Sons, 1969.

Kappes, B. M., 'Sequence effects of relaxation training, EMG, and temperature biofeedback on anxiety, symptom report, and self-concept', *Journal of Clinical Psychology*, 1983; 39 (2): 203–8.

Kashulin, P. A. et al., 'Phenolic biochemical pathway in plants can be used for the bioindication of heliogeophysical factors', 'Physics of Auroral Phenomena', Proceedings, XXV Annual Seminar, Apatity: Polar Geophysical Institute, Kola Science Center, Russian Academy of Science, February 26–March 1; 2002: 153–6.

Kaufman, M., 'Meditation gives brain a charge, study finds', *Washington Post*, January 3, 2005.

Keen, J., *Consciousness, Intent and the Structure of the Universe*, Victoria, BC: Trafford Publishing, 2005.

Keicolt-Glaser, J. K., 'Hostile marital interactions, proinflammatory cytokine production, and wound healing', *Archives of General Psychiatry*, 2005; 62 (12): 1377–84.

Koren, S. A. and Persinger, M. A., 'Possible disruption of remote viewing by complex weak magnetic fields around the stimulus site and the possibility of accessing real phase space: a pilot study', *Perceptual and Motor Skills*, 2002; 95 (3 Pt 1): 989–98.

Korotkov, K. et al., 'Assessing biophysical energy transfer mechanisms in living systems: the basis of life processes', *Journal of Alternative and Complementary Medicine*, 2004; 10 (1): 49–57.

Korotkov, K. et al., 'Stress diagnosis and monitoring with new computerized "Crown-TV" device', *Journal of Pathophysiology*, 1998; 5: 227.

Kosslyn, S. M. et al., 'Hypnotic visual illusion alters color processing in the brain', *American Journal of Psychiatry*, 2000; 157: 1279–84.

Krippner, S., 'Dancing with the trickster: notes for a transpersonal autobiography', *International Journal of Transpersonal Studies*, 2002; 21: 1–18.

Krippner, S., 'Possible geomagnetic field effects in psi phenomena.' Paper presented at international parapsychology conference in Recife, Brazil, November 1997.

Krippner, S., 'Psi research and the human brain's "reserve capacities"', *Dynamical Psychology*, 1996; available online: http://goertzel.org/dynapsych/1996/stan.html.

Krippner, S., 'Psychoneurological dimensions of anomalous experience in relation to religious belief and spiritual practice', in K. Bulkeley (ed.), *Soul, Psyche, Brain*, New York: Palgrave Macmillan, 2005: 61–92.

Krippner, S., 'Stigmatic phenomenon: an alleged case in Brazil', *Journal of Scientific Exploration*, 2002; 16 (2): 207–24.

Krippner, S., 'The epistemology and technologies of shamanic states of consciousness', *Journal of Consciousness Studies*, 2000; 7: 93–118.

Krippner, S., 'The technologies of shamanic states of consciousness', in M. Schlitz et al. (eds.), *Consciousness and Healing: Integral Approaches to Mind–Body Medicine*, St Louis, Mo.: Elsevier Churchill Livingstone, 2005: 376–90.

Krippner, S., 'Trance and the trickster: hypnosis as a liminal phenomenon', *International Journal of Clinical and Experimental Hypnosis*, 2005, 53 (2): 97–118.

Krippner, S. and Persinger, M., 'Evidence for enhanced congruence between dreams and distant target material during periods of decreased geomagnetic activity', *Journal of Scientific Exploration*, 1996; 10: 487–93.

Krippner, S. et al., 'Geomagnetic factors in subjective precognitive dream experiences', *Journal of the Society for Psychical Research*, 2000; 64 (859): 109–18.

Krippner, S. et al., 'Physiological and geomagnetic correlates of apparent anomalous phenomena observed in the presence of a Brazilian "sensitive"', *Journal of Scientific Exploration*, 1996; 10: 281–98.

Krippner, S. et al., 'The indigenous healing tradition in Calabria, Italy.' Paper presented at the Annual Conference for the Study of Shamanism and Alternative Modes of Healing, San Rafael, California, September 2004.

Krippner, S. et al., 'The Ramtha phenomenon: psychological, phenomenological, and geomagnetic data', *Journal of the American Society for Psychical Research*, 1998; 92: 1–24.

Krippner, S. et al., 'Working with Ramtha: is it a "high risk" procedure?' *Proceedings of Presented Papers*, the Parapsychological Association 41st Annual Convention, 1998: 50–63.

Krucoff, M. et al., 'From efficacy to safety concerns: a STEP forward or a step back for clinical research and intercessory prayer? The Study of Therapeutic Effects of Intercessory Prayer (STEP)', *American Heart Journal*, 2006; 151 (4): 762.

Krucoff, M. et al., 'Music, imagery, touch and prayer as adjuncts to interventional cardiac care: the Monitoring and Actualisation of Noetic Trainings (MANTRA) II randomised study', *The Lancet*, 2005; 366: 211–17.

Krucoff, M. W., 'Integrative noetic therapies as adjuncts to percutaneous intervention during unstable coronary syndromes: Monitoring and Actualization of Noetic Training (MANTRA) feasibility pilot', *American Heart Journal*, 2001; 142 (5): 760–7.

Krueger, A. P. and Sobel, D. S., 'Air ions and health', in David S. Sobel (ed.), *Ways of Health: Holistic Approaches to Ancient and Contemporary Medicine*, New York: Harcourt Brace Jovanovich, 1979.

Larina, O. N. et al., 'Effects of spaceflight factors on recombinant protein expression in *E. coli* producing strains', in 'Biomedical Research on the Science/NASA Project', Abstracts of the Third US/Russian Symposium, Huntsville, Alabama, November 10–13, 1997: 110–11.

Lashley, J. K. et al., 'An empirical account of temperature biofeedback applied in groups', *Psychological Reports*, 1987; 60 (2): 379–88.

Laszlo, E., *Science and the Akashic Field: An Integral Theory of Everything*, Rochester, Vt.: Inner Traditions, 2004.

Laszlo, E., *The Interconnected Universe: Conceptual Foundations of Transdiscipinary Unified Theory*, Singapore: World Scientific Publishing, 1995.

Lazar, S. et al., 'Meditation experience is associated with increased cortical thickness', *NeuroReport*, 2005; 16: 1893–7.

Lazar, S. W. et al., 'Functional brain mapping of the relaxation response and meditation', *NeuroReport*, 2000; 11: 1581–5.

Leibovici, L., 'Alternative (complementary) medicine: a cuckoo in the nest of empiricist reed warblers', *British Medical Journal*, 1999; 319: 1629–32.

Leibovici, L., 'Effects of remote, retroactive intercessory prayer on outcomes in patients with blood stream infection: randomized controlled trial', *British Medical Journal*, 2001; 323 (7327): 1450–1.

LeShan, L., *The Medium, the Mystic and the Physicist: Towards a Theory of the Paranormal*, New York: Helios, 2003.

LeShan L. L. and Gassmann, M. L., 'Some observations on psychotherapy with patients with neoplastic disease', *American Journal of Psychotherapy*, 1958; 12: 723–34.

Letters, 'Effect of retroactive prayer', *British Medical Journal*, 2002; 324: 1037.

Letters, BMJ Online, December 22, 2003.

Lobach, E. and Bierman, D. J., 'Who's calling at this hour? Local sidereal time and telephone telepathy', Proceedings of Presented Papers, 47th Annual Convention of the Parapsychological Association Convention, Vienna, August 5–8, 2004.

Luskin, F. M. et al., 'A review of mind-body therapies in the treatment of cardiovascular disease, Part 1: Implications for the elderly', *Alternative Therapies in Health and Medicine*, 1998; 4 (3): 46–61.

Luskin, F. M. et al., 'A review of mind/body therapies in the treatment of musculoskeletal disorders with implications for the elderly', *Alternative Therapies in Health and Medicine*, 2000; 6 (2): 46–56.

Lutz, A. et al., 'Long-term meditators self-induce high-amplitude gamma synchrony during mental practice', *Proceedings of the National Academy of Science*, 2004; 16, 101(46): 16369–73.

McCraty, R., 'Influence of cardiac afferent input on heart-brain synchronization and cognitive performance', *International Journal of Psychophysiology*, 2002; 45 (1–2): 72–3.

McCraty, R. et al., 'Electrophysiological evidence of intuition: Part 1. The surprising role of the heart', *Journal of Alternative and Complementary Medicine*, 2004; 10 (1): 133–43.

McCraty, R. et al., 'Electrophysiological evidence of intuition: Part 2. A system-wide process?' *Journal of Alternative and Complementary Medicine*, 2004; 10 (2): 325–36.

McCraty, R. et al., 'Head-heart entrainment: a preliminary survey', in Proceedings of the Brain–Mind Applied Neurophysiology EEG Neurofeedback Meeting. Key West, Florida, 1996.

McCraty, R. et al., 'The electricity of touch: detection and measurement of cardiac energy exchange between people', in Karl H. Pribram (ed.), *Brain and Values: Is a Biological Science of Values Possible?* Mahwah, NJ: Lawrence Erlbaum Associates, 1998: 359–79.

McGugan, E. A., 'Sudden unexpected deaths in epileptics – literature review', *Scottish Medical Journal*, 1999; 44 (5): 137–9.

McKay, B. and Persinger, M., 'Geophysical Variables and Behavior, LXXXVII: Effects of synthetic and natural geomagnetic patterns on maze learning', *Perceptual and Motor Skills*, 1999; 89 (3 pt 1): 1023–4.

McTaggart, L., *The Field: The Quest for the Secret Force of the Universe*, London: HarperCollins, 2001.

McTaggart, L., *What Doctors Don't Tell You: The Truth about the Dangers of Modern Medicine*, London: HarperCollins, 2005.

Mailer, N., *The Fight*, London and New York: Penguin, 2000.

Malle, B. F. et al., *Intentions and Intentionality: Foundations of Social Cognition*, Cambridge, Mass.: The MIT Press, 2001.

Maris, G. et al., 'Geomagnetic consequences of the solar flares during the last Hale solar cycle (II)', in H. Sawaya-Lacoste (ed.), *Proceedings of the Second Solar Cycle and Space Weather Euroconference, September 24–29, 2001, Vico Equense, Italy*. Noordwijk, the Netherlands: ESA Publications, 2002: 451–4.

Michon, A. L. and Persinger, M. A., 'Experimental simulation of the effects of increased geomagnetic activity upon nocturnal seizures in epileptic rats', *Neuroscience Letters*, 1997; 224: 53–6.

Michon, A. L. et al., 'Attempts to simulate the association between geomagnetic activity and spontaneous seizures in rats using experimentally generated magnetic fields', *Perceptual and Motor Skills*, 1996; 82 (2): 619–26.

Mikulecky, M., 'Lunisolar tidal waves, geomagnetic activity and epilepsy in the light of multivariate coherence', *Brazilian Journal of Medicine*, 1996; 29 (8): 1069–72.

Miller, R. N., 'Study of remote mental healing', *Medical Hypotheses*, 1982; 8: 481–90.

Miller, R. N., 'The positive effect of prayer on plants', *Psychic*, 1972; 3 (5): 24–5.

Minas, S. C., 'Mental practice of a complex perceptual-motor skill', *Journal of Human Movement Studies*, 1978; 4: 102–7.

Mizun, Y. G. and Mizun, P. G., *Space and Health*, Moscow: 'Znanie', 1984.

Monafo, W. W. and West, M. A., 'Current recommendations for topical burn therapy', *Drugs*, 1990; 40: 364–73.

Moseley, J. B. et al., 'A controlled trial of arthroscopic surgery for osteoarthritis of the knee', *New England Journal of Medicine*, 2002; 347: 81–8.

Mourou, G. A. and Umstadter, D., 'Extreme light', in 'The edge of physics', Special edition of *Scientific American*, 2002; 286: 80–6.

Muehsam, D. J. et al., 'Effects of Qigong on cell-free myosin phosphorylation: preliminary experiments', *Subtle Energies and Energy Medicine*, 1994; 5 (1): 93–108.

Mumford, B. and Hall, C., 'The effects of internal and external imagery on performing figures in figure skating', *Canadian Journal of Applied Sport Sciences*, 1985; 10: 171–7.

Murphy, M. et al., *The Physiological and Psychological Effects of Meditation: A Review of Contemporary Research With a Comprehensive Bibliography, 1931–1996*, Petaluma, Calif.: The Institute of Noetic Sciences, 1997.

Nash, C. B., 'Test of psychokinetic control of bacterial mutation', *Journal of the American Society for Psychical Research*, 1984; 78: 145–52.

Nash, C. B. and Nash, C. S., 'The effect of paranormally conditioned solution on yeast fermentation', *Journal of Parapsychology*, 1967; 31: 314.

Nelson, L. and Schwartz, G. E., 'Human biofield and intention detection: individual differences', *Journal of Alternative and Complementary Medicine*, 2005; 11 (1): 93–101.

Nelson, R., 'Correlation of global events with REG data: an internet-based, nonlocal anomalies experiment', *Journal of Parapsychology*, 2001; 65: 247–71.

Nelson, R., 'Wishing for good weather: a natural experiment in group consciousness', *Journal of Scientific Exploration*, 1997; 11 (1): 47–58.

Nelson, R. D., 'Coherent consciousness and reduced randomness: correlations on September 11, 2001', *Journal of Scientific Exploration*, 2002; 16 (4): 549–70.

Nelson, R. D., 'The physical basis of intentional healing systems', Princeton Engineering Anomalies Research, PEAR Technical Report, 99001, Princeton, New Jersey, January 1999.

Nelson, R. D. et al., 'Correlation of continuous random data with major world events', *Foundations of Physics Letters*, 2002; 15 (6): 537–50.

Nelson, R. D. et al., 'FieldREG anomalies in group situations', *Journal of Scientific Exploration*, 1996; 10 (1): 111–14.

Nelson, R. D. et al., 'FieldREGII: consciousness field effects: replications and explorations', *Journal of Scientific Exploration*, 1998; 12 (3): 425–54.

Novikova, K. F. and Ryvkin, B. A., 'Solar activity and cardiovascular diseases', in Gnevyshev, M. N. and Ol, A. I. (eds.), *Effects of Solar Activity on the Earth's Atmosphere and Biosphere*, Academy of Science, USSR (translated from the Russian). Jerusalem: Israel Program for Scientific Translations, 1977: 184–200.

O'Connor, R. P. and Persinger, M. A., 'Geophysical Variables and Behavior, LXXXII: A strong association between sudden infant death syndrome (SIDS) and increments of global geomagnetic activity – possible support for the melatonin hypothesis', *Perceptual and Motor Skills*, 1997; 84: 395–402.

O'Connor, R. P. and Persinger, M. A., 'Geophysical Variables and Behavior, LXXXV: Sudden infant death syndrome, bands of geomagnetic activity and pc1 (0.2 to 4 HZ) geomagnetic micropulsations', *Perceptual and Motor Skills*, 1999; 88: 391–7.

O'Laoire, S., 'An experimental study of the effects of distant, intercessory prayer on self-esteem, anxiety and depression', *Alternative Therapies in Health and Medicine*, 1997; 3 (6): 19–53.

Olendzki, A., 'The fourth foundation of mindfulness', *Insight Journal*, Spring 2004: 13–17.

Oliver, R. T. D., 'Surveillance as a possible option for management of metastic renal cell carcinoma', *Seminars in Urology*, 1989; 7: 149–52.

Oraevskii, V. N. et al., 'An influence of geomagnetic activity on the functional status of the body', *Biofizika*, 1998; 43 (5): 819–26.

Oraevskii, V. N. et al., 'Medico-biological effect of natural electromagnetic variations', *Biofizika*, 1998; 43 (5): 844–8.

O'Regan, B. and Hirshberg, C., *Spontaneous Remission: An Annotated Bibliography*, Petaluma, Calif.: The Institute of Noetic Sciences, 1993.

Orme-Johnson, W. et al., 'International peace project in the Middle East: the effects of the Maharishi technology of the unified field', *Journal of Conflict Resolution*, 1988; 32: 776–812.

Oshansky, B. and Dossey, L., 'Comments on responses to "retroactive prayer: a preposterous hypothesis?"' *British Medical Journal*, 2003; 327: 1465–8.

Oshansky, B. and Dossey, L., 'Retroactive prayer: a preposterous hypothesis?' *British Medical Journal*, 2003; 327: 20–7.

Paivio, A., 'Cognitive and motivational functions of imagery in human performance', *Canadian Journal of Applied Sport Sciences*, 1985; 10 (4): 22S–28S.

Pates, J. and Maynard, I., 'Effects of hypnosis on flow states and golf performance', *Perceptual and Motor Skills*, 2000; 9: 1057–75.

Pates, J. et al., 'The effects of hypnosis on flow states and three-point shooting in basketball players', *The Sport Psychologist*, 2002; 16: 34–47.

Peniston E. and Kulkosky, P. J., 'Alcoholic personality and alpha-theta brainwave training', *Medical Psychotherapy*, 1990; 3: 37–55.

Peniston, E. and Kulkosky, P. J., 'Alpha-theta brainwave training and beta-endorphin levels in alcoholics', *Alcoholism: Clinical and Experimental Research*, 1989; 13: 271–9.

Peoc'h, R., 'Chicken imprinting and the tychoscope: an Anpsi experiment', *Journal of the Society for Psychical Research*, 1988; 55: 1

Peoc'h, R., 'Psychokinesis experiments with human and animal subjects upon a robot moving at random', *Journal of Parapsychology*, September 1, 2002.

Peoc'h, R., 'Psychokinetic action of young chicks on the path of an "illuminated source"', *Journal of Scientific Exploration*, 1995; 9 (2): 223.

Persinger, M. A., 'ELF field meditation in spontaneous psi events. Direct information transfer or conditioned elicitation?' *Psychoenergetic Systems*, 1975; 3: 155–69.

Persinger, M. A., 'Enhancement of images of possible memories of others during exposure to circumcerebral magnetic fields: correlations with ambient geomagnetic activity', *Perceptual and Motor Skills*, 2002; 95 (2): 531–43.

Persinger, M. A., 'Geophysical Variables and Behavior, XXX: Intense paranormal activities occur during days of quiet global geomagnetic activity', *Perceptual and Motor Skills*, 1985; 61: 320–2.

Persinger, M.A., 'Increased emergence of alpha activity over the left but not the right temporal lobe within a dark acoustic chamber: differential response of the left but not the right hemisphere to transcerebral magnetic fields', *International Journal of Psychophysiology*, 1999; 34 (2): 163–9.

Persinger, M. A., 'Sudden unexpected death in epileptics following sudden, intense, increases in geomagnetic activity: prevalence of effect and potential mechanisms', *International Journal of Biometeorology*, 1995; 38: 180–7.

Persinger, M. A. and Koren, S. A., 'Experiences of spiritual visitation and impregnation: potential induction by frequency-modulated transients from an adjacent clock', *Perceptual and Motor Skills*, 2001; 92 (1): 35–6.

Persinger, M. A. and Krippner, S., 'Dream ESP experiments and geomagnetic activity', *Journal of the American Society for Psychical Research*, 1989; 83: 101–16.

Persinger, M. A. et al., 'Differential entrainment of electroencephalographic activity by weak complex electromagnetic fields', *Perceptual and Motor Skills*, 1997; 84 (2): 527–36.

Persinger, M. A. et al., 'Remote viewing with the artist Ingo Swann: neuropsychological profile, electroencephalographic correlates, magnetic resonance imaging (MRI), and possible mechanisms', *Perceptual and Motor Skills*, 2002; 94: 927–9.

Petro, V. M. et al., 'An influence of changes of magnetic field of the earth on the functional state of humans in the conditions of space mission', Proceedings, International Symposium 'Computer Electro-Cardiograph on Boundary of Centuries', Moscow, Russian Federation, 27–30 April, 1999.

Popp, F.-A., 'Evidence of non-classical (squeezed) light in biological systems', *Physics Letters A*, 2002; 293 (1–2): 98–102.

Popp, F.-A. et al., 'Further analysis of delayed luminescence of plants', *Journal of Photochemistry and Photobiology B: Biology*, 2005; 78: 235–44.

Popp, F.-A. et al., 'Mechanism of interaction between electromagnetic fields and living organisms', *Science in China* (Series C), 2000; 43 (5): 507–18.

Popp, F.-A. et al., 'Nonsubstantial biocommunication in terms of Dicke's Theory', in M. W. Ho, F.-A. Popp and U. Warnke (eds.), *Bioelectrodynamics and Biocommunication*, Singapore: World Scientific Publishing, 1994: 293–317.

Puthoff, H. E., 'Ground state of hydrogen as a zero-point-fluctuation-determined state', *Physical Review D*, 1987; 35: 3266.

Pyatnitsky, L. N. and Fonkin, V. A., 'Human consciousness influence on water structure', *Journal of Scientific Exploration*, 1995; 9 (1): 89.

Radin, D. I., 'A dog that seems to know when his owner is coming home: effect of environmental variables', *Journal of Scientific Exploration*, 2002; 16 (45): 579–92.

Radin, D. I., 'A dog that seems to know when his owner is coming home: effects of geomagnetism and local sidereal time', Boundary Institute Technical Report.

Radin, D. I., 'Beyond belief: exploring interaction among body and environment', *Subtle Energies and Energy Medicine*, 1992; 2 (3): 1–40.

Radin, D. I., *Entangled Minds*, New York: Paraview, 2006.

Radin, D. I., *The Conscious Universe*, London: HarperCollins, 1997.

Radin, D. I., 'Time-reversed human experience: experimental evidence and implications', *Journal of Nonlocality and Remote Mental Interactions*, July 2003; II (2).

Radin, D. I. and Nelson, R., 'Evidence for consciousness-related anomalies in random physical systems', *Foundations of Physics*, 1989; 19 (12): 1499–514.

Radin, D. I. et al., 'Effects of distant healing intention through time and space: two exploratory studies', *Proceedings of Presented Papers*. The 41st Annual Convention of the Parapsychological Association, Halifax, Nova Scotia: Parapsychological Association, 1998: 143–61.

Radin, D. I. et al., 'Effects of healing intention on cultured cells and truly random events', *Journal of Alternative and Complementary Medicine*, 2004; 10: 103–12.

Radin, D. I. et al., 'Effects of motivated distant intention on electrodermal activity.' Paper presented at the annual conference of the Parapsychological Association, Stockholm, Sweden, August 2006.

Radin, D. I., 'Environmental modulation and statistical equilibrium in mind-matter interaction', *Subtle Energies and Energy Medicine*, 1993; 4 (1): 1–30.

Radin, D. I., 'Event-related electroencephalographic correlations between isolated human subjects', *Journal of Alternative and Complementary Medicine*, 2004; 10 (2): 315–23.

Radin, D. I., 'Evidence for relationship between geomagnetic field fluctuations and skilled physical performance.' Paper presented at the 11th Annual Meeting of the Society for Scientific Explorations, Princeton, New Jersey, June 1992.

Radin, D. I., 'Exploring relationships between random physical events and mass human attention: asking for whom the bell tolls', *Journal of Scientific Exploration*, 2002; 16 (4): 533–47.

Radin, D. I., 'For whom the bell tolls; a question of global consciousness', *Noetic Sciences Review*, 2003; 63: 8–13 and 44–5.

Radin, D. I., 'Geomagnetic field fluctuations and sports performance', *Subtle Energies and Energy Medicine*, 1996; 6 (3): 217–26.

Radin, D. I., 'Unconscious perception of future emotions: an experiment in presentiment', *Journal of Scientific Exploration*, 1997; 11 (2): 163–80.

Radin, D. I. and May, E. C., 'Evidence for a retrocausal effect in the human nervous system', Boundary Institute Technical Report 2000–1.

Radin, D. I. and Rebman, J. M., 'Seeking psi in the casino', *Journal of the Society for Psychical Research*, 1998; 62 (850): 193–219.

Radin, D. I. and Schlitz, M. J. 'Gut feelings, intuition, and emotions: an exploratory study', *Journal of Alternative and Complementary Medicine*, 2005; 11 (5): 85–91.

Radin, D. I. and Utts, J. M., 'Experiments investigating the influence of intention on random and pseudorandom events', *Journal of Scientific Exploration*, 1989; 3: 65–79.

Radin, D. I., Taylor, R. D. and Braud, W., 'Remote mental influence of human electrodermal activity: a pilot replication', *European Journal of Parapsychology*, 1995; 11: 19–34.

Radin, D. I. et al., 'Geomagnetism and psi in the ganzfeld', *Journal of the Society for Psychical Research*, 1994; 59 (834): 352–63.

Ranganathan, V. K. et al., 'Increasing muscle strength by training the central nervous system without physical exercise', *Society for Neuroscience Abstracts*, 2001; 31: 17.

Ranganathan, V. K. et al., 'Level of mental effort determines training-induced strength increases', *Society of Neuroscience Abstracts*, 2002; 32: 768.

Raps, A. et al., 'Geophysical Variables and Behavior, LXIX: Solar activity and admission of psychiatric inpatients', *Perceptual and Motor Skills*, 1992; 74: 449.

Raud, P. C., 'Psychospiritual dimensions of extraordinary survival', *Journal of Humanistic Psychology*, 1989; 29: 59–83.

Raynes, B., 'Interview with Todd Murphy', *Alternative Perceptions Magazine* online, April 2004; No 78.

Reece, K. et al., 'Positive well-being changes associated with giving and receiving Johrei healing', *Journal of Alternative and Complementary Medicine*; 2005, 11 (3): 455–7.

Rein, G., 'Biological effects of quantum fields and their role in the natural healing process', *Frontier Perspectives*, 1998; 7: 16–23.

Rein, G., 'Effect of conscious intention on human DNA'. Paper presented at the International Forum on New Science, Denver, Colorado, October 1996.

Rein, G. and McCraty, R., 'Structural changes in water and DNA associated with new physiologically measurable states', *Journal of Scientific Exploration*, 1994; 8 (3): 438–9.

Resch, J. et al., 'Distributing entanglement and single photons through an intra-city, free-space quantum channel', *Optics Express*, 2005; 13 (1): 202–9.

Reznik, B., 'Entanglement from the vacuum', *Foundations of Physics*, 2003; 33: 167–76.

Richards, T. et al., 'Replicable functional magnetic resonance imaging evidence of correlated brain signals between physically and sensory isolated subjects', *Journal of Alternative and Complementary Medicine*, 2005; 11 (6): 955–63.

Rinpoche, S., *The Tibetan Book of Living and Dying*, San Francisco: HarperSanFrancisco, 1994.

Roney-Dougal, S. M. and Solfvin, J., 'Field study of an enhancement effect on lettuce seeds – replication study', *Journal of Parapsychology*, 2003; 67 (2): 279–98.

Rose, G. D. et al., 'The behavioral treatment of Raynaud's disease: a review', *Biofeedback and Self Regulation*, 1987; 12: 257–72.

Rosenblum, B. and Kuttner, F., 'The observer in the quantum experiment', *Foundations of Physics*, 2002; 32 (8): 1273–93.

Rotella, R. J. et al., 'Cognitions and coping strategies of elite skiers: an exploratory study of young developing athletes', *Journal of Sport Psychology*, 1980; 2: 350–4.

Rubik, B. et al., 'In vitro effect of Reiki treatment on bacterial cultures: role of experimental context and practitioner well-being', *Journal of Alternative and Complementary Medicine*, 2006; 12 (1): 7–13.

Rushall, B. S., 'Covert modeling as a procedure for altering an elite athlete's psychological state', *Sport Psychologist*, 1988; 2: 131–40.

Rushall, B. S., 'The restoration of performance capacity by cognitive restructuring and covert positive reinforcement in an elite athlete', in J. R. Cautela and A. J. Kearney (eds.), *Covert Conditioning Casebook*, Boston, Mass.: Thomson Brooks/Cole, 1993.

Rushall, B. S. and Lippman, L. G., 'The role of imagery in physical performance', *International Journal for Sport Psychology*, 1997; 29: 57–72.

Salmon, J. et al., 'The use of imagery by soccer players', *Journal of Applied Sport Psychology*, 1994; 6: 116–33.

Sancier, K. M., 'Electrodermal measurements for monitoring the effects of a Qigong workshop', *Journal of Alternative and Complementary Medicine*, 2003; 9 (2): 235–41.

Sancier, K. M., 'Medical applications of *Qigong* and emitted *Qi* on humans, animals, cell cultures, and plants: review of selected scientific research', *American Journal of Acupuncture*, 1991; 19 (4): 367–77.

Sancier, K. M., 'Search for medical applications of Qigong with the computerized Qigong Database™', *Journal of Alternative and Complementary Medicine*, 2001; 7 (1): 93–5.

Satinsky, D., 'Biofeedback treatment for headache: a two-year follow-up study', *American Journal of Clinical Biofeedback*, 1981; 4 (1): 62–5.

Scargle, J. D., 'Commentary: Was there evidence of global consciousness on September 11, 2001?' *Journal of Scientific Exploration*, 2002; 16 (4): 571–7.

Schlitz, M., 'Can science study prayer?' *Shift: At the Frontiers of Consciousness*, 2006, September–November; 12: 38–9.

Schlitz, M., 'Distant healing intention: definitions and evolving guidelines for laboratory studies', *Alternative Therapies in Health and Medicine*, 2003; 9 (3 Suppl): A31–43.

Schlitz, M., 'Intentionality in healing: mapping the integration of body, mind, and spirit', *Alternative Therapies in Health and Medicine*, 1995; 1 (5): 119–20.

Schlitz, M. et al. (eds.), *Consciousness and Healing: Integral Approaches to Mind–Body Healing*, St. Louis, Mo.: Elsevier Churchill Livingstone, 2005.

Schlitz, M. J. and Braud, W. G., 'A methodology for the objective study of transpersonal imagery', *Journal of Scientific Exploration*, 1989; 3 (1): 43–63.

Schlitz, M. J. and Braud, W. G., 'Distant intentionality and healing: assessing the evidence', *Alternative Therapies in Health and Medicine*, 1997; 3 (6): 62–73.

Schlitz, M. J. and Honorton, C., 'Ganzfeld psi performance within an artistically gifted population', *Journal of the American Society for Psychical Research*, 1992; 86 (2): 83–98.

Schlitz, M. J. and LaBerge, S., 'Autonomic detection of remote observation; two conceptual replications', in D. Bierman (ed.), *Proceedings of Presented Papers: 37th Annual Parapsychological Association Convention, Amsterdam*. Fairhaven, Mass.: Parapsychological Association, 1994: 465–78.

Schmidt, H., 'Additional effect for PK on pre-recorded targets', *Journal of Parapsychology*, 1985; 49: 229–44.

Schmidt, H., 'PK tests with and without preobservation by animals', in L. S. Henkel and J. Palmer (eds.), *Research in Parapsychology, 1989*, Metuchen, NJ: Scarecrow Press, 1990: 15–19.

Schmidt, H., 'Random generators and living systems as targets in retro-PK experiments', *Journal of the American Society for Psychical Research*, 1997; 912 (1): 1–13.

Schmidt, H. and Stapp, H., 'PK with prerecorded random events and the effects of preobservation', *Journal of Parapsychology*, 1993; 57 (4): 331–49.

Schmidt, H. and Stapp, H., 'Study of PK with prerecorded random events and the effects of preobservation', *Journal of Parapsychology*, 1993; 57: 351.

Schmidt, S. et al., 'Distant intentionality and the feeling of being stared at: two meta-analysis', *British Journal of Psychology*, 2004; 95: 235–47.

Schoenberger, N. E. et al., 'Flexyx neurotherapy system in the treatment of traumatic brain injury: an initial evaluation', *Journal of Head Trauma Rehabilitation*, 2001; 16 (3): 260–74.

Schwartz, G. and Russek, L., 'Subtle energies – electrostatic body motion registration and the human antenna-receiver effect: a new method for investigating interpersonal dynamical energy system interactions', *Subtle Energies and Energy Medicine*, 1996; 7 (2): 149–84.

Schwartz, G. E., 'Biofeedback, self-regulation, and the patterning of physiological processes', *American Scientist*, 1975; 63 (3): 314–24.

Schwartz, G. E. and Russek, L. G., 'Dynamical energy systems and modern physics', *Alternative Therapies in Health and Medicine*, 1997; 3: 46–56.

Schwartz, G. E. et al., 'Interpersonal hand-energy registration: evidence for implicit performance and perception', *Subtle Energies and Energy Medicine*, 1995; 6 (3): 183–200.

Schwartz, S. A. et al., 'Infrared spectra alteration in water proximate to the palms of therapeutic practitioners', *Subtle Energies and Energy Medicine*, 1991; 1: 43–57.

Scott, W. B., 'To the stars', *Aviation Week and Space Technology*, March 4, 2004: 50–3.

Semikhina, L. P. and Kiselev, V. P., 'Effect of weak magnetic fields on the properties of water and ice', *Zabedenii, Fizika*, 1988; 5: 13–17.

Seto, A. et al., 'Detection of extraordinary large biomagnetic field strength from the human hand during external qi emission', *Acupuncture and Electrotherapeutics Research International*, 1992; 17: 75–94.

Sheldrake, R., *Dogs that Know When Their Owners Are Coming Home and Other Unexplained Powers of Animals*, London: Three Rivers Press, 2000.

Sheldrake, R., *The Sense of Being Stared At and Other Aspects of The Extended Mind*, London: Hutchinson, 2003.

Sherwood, S. J. and Roe, C. A., 'A review of dream ESP studies conducted since the Maimonides dream ESP programme', *Journal of Consciousness Studies*, 2003; 10: 85–109.

Sicher, F. et al., 'A randomized double-blind study of the effect of distant healing in a population with advanced AIDS: report of a small scale study', *Western Journal of Medicine*, 1998; 168 (6): 356–63.

Siegel, B., *Love, Medicine and Miracles: Lessons Learned about Self-Healing from a Surgeon's Experience with Exceptional Patients*, London: HarperCollins, 1990.

Silver, B. V. et al., 'Temperature biofeedback and relaxation training in the treatment of migraine headaches: one-year follow-up', *Biofeedback and Self Regulation*, 1979; 4 (4): 359–66.

Simonton, C. O. et al., *Getting Well Again*, New York: Bantam, 1980.

Simpson, S. H. et al., 'A meta-analysis of the association between adherence to drug therapy and mortality', *British Medical Journal*, 2006; 333: 15–19.

Singer, W., 'Neuronal synchrony: a versatile code for the definition of relations?' *Neuron*, 1999; 24: 49–65.

Smith, C. W., 'Is a living system a macroscopic quantum system?' *Frontier Perspectives*, 1998; 7 (1) Fall/Winter: 9.

Smith, D. et al., 'The effect of mental practice on muscle strength and EMG activity', *Proceedings of the British Psychological Society Annual Conference*, 1998; 6 (2): 116.

Snel, F. W. J. J. and van der Sijde, P. C., 'The effect of retro-active distance healing on *Babeia rodhani* (rodent malaria) in rats', *European Journal of Parapsychology*, 1990; 8: 123–30.

Sorensen, A. et al., 'Many-particle entanglement with Bose–Einstein condensates', *Nature*, 2001; 409 (6816): 63–6.

Spottiswoode, J., 'Geomagnetic fluctuations and free response anomalous cognition: a new understanding', *Journal of Parapsychology*, 1997; 61: 3–12.

Spottiswoode, J. P., 'Effect of ambient magnetic field fluctuations on performance in a free response anomalous cognition task: a pilot study', *Proceedings of the 36th Annual Convention of the Parapsychological Association*, 1993: 143–56.

Spottiswoode, J. P. and May, E. C., 'Anomalous cognition effect size: dependence on sidereal time and solar wind parameters', Spottiswoode library, http://www.jsasoc.com/library.html.

Spottiswoode, S. J. P., 'Apparent association between effect size in free response anomalous cognition experiments and local sidereal time', *Journal of Scientific Exploration*, 1997: 11 (2): 109–22.

Spottiswoode, S. J. P. and May, E., 'Evidence that free response anomalous cognitive performance depends upon local sidereal time and geomagnetic fluctuations', *Presentation Abstracts, Sixteenth Annual Meeting of the Society for Scientific Exploration*, June 1997: 8.

Squires, E. J., 'Many views of one world – an interpretation of quantum theory', *European Journal of Physics*, 1987; 8: 173.

Standish, L. J. et al., 'Electroencephalographic evidence of correlated event-related signals between the brains of spatially and sensory isolated human subjects', *Journal of Alternative and Complementary Medicine*, 2004; 10 (2): 307–14.

Standish, L. J. et al., 'Evidence of correlated functional magnetic resonance imaging signals between distant human brains', *Alternative Therapies in Health and Medicine*, 2003; 9 (1): 122–5.

Stapp, H. P., 'A bell-type theorem without hidden variables', *American Journal of Physics*, 2004; 72: 30–3.

Stapp, H. P., 'Theoretical model of a purported empirical violation of the predictions of quantum theory', *Physical Review A*, 1994; 50 (1): 18–22.

Stein, J., 'Just say Om', *Time* magazine, August 4, 2003: 49–56.

Sterman, M. B., 'Basic concepts and clinical findings in the treatment of seizure disorders with EEG operant conditioning', *Clinical Electroencephalography*, 2000; 31(1): 45–55.

Sterman, M. B., 'Epilepsy and its treatment with EEG feedback therapy', *Annals of Behavioral Medicine*, 1986; 8: 21–5.

Sterman, M. B., 'Neurophysiological and clinical studies of sensorimotor EEG biofeedback training: some effects on epilepsy', *Seminars in Psychiatry*, 1973; 5 (4): 507–25.

Sterman, M. B., 'The challenge of EEG biofeedback in the treatment of epilepsy: a view from the trenches', *Biofeedback*, 1997; 25 (1): 6–7.

Stibor, A. et al., 'Talbot-Lau interferometry with fullerenes: sensitivity to inertial forces and vibrational dephasing', *Laser Physics*, 2005; 15: 10–17.

Stoilova, I. and Zdravev, T., 'Influence of the geomagnetic activity on the human functional systems', *Journal of the Balkan Geophysical Society*, 2000; 3 (4): 73–6.

Stone, J., Course Handbook: Training in Compassionate-Loving Intention (unpublished), 2003.

Stone, J., 'Effects of a compassionate/loving intention as a therapeutic intervention by partners of cancer patients: a randomized controlled feasibility study', in press.

Stoupel, E., 'Relationship between suicide and myocardial infarction with regard to changing physical environmental conditions', *International Journal of Biometeorology*, 1994; 38 (4): 199–203.

Stoupel, E. et al., 'Clinical cosmobiology: the Lithuanian study, 1990–1992', *International Journal of Biometeorology*, 1995; 38: 204–8.

Stoupel, E. et al., 'Suicide-homicide temporal interrelationship, links with other fatalities and environmental physical activity', *Crisis*, 2005; 26: 85–9.

Suinn, R. M., 'Imagery rehearsal applications to performance enhancement', *The Behavior Therapist*, 1985; 8: 155–9.

Surwillo, W. W. and Hobson, D. P., 'Brain electrical activity during prayer', *Psychological Reports*, 1978; 43 (1): 135–43.

Swets, J. A. and Bjork, R. A., 'Enhancing human performance: an evaluation of "New Age" techniques considered by the U.S. Army', *Psychological Science*, 1990; 1: 85–96.

Talbot, M., *Mysticism and the New Physics*, London: Penguin, 1993.

Targ, E., 'Research methodology for studies of prayer and distant healing', *Complementary Therapies in Nursing and Midwifery*, 2002; 8: 29–41.

Tart, C., 'Physiological correlates of psi cognition', *International Journal of Parapsychology*, 1963; 5: 375–86.

Tart, C. T., *Body Mind Spirit: Exploring the Parapsychology of Spirituality*, Charlottesville, Va.: Hampton Roads Publishing Company, 1997.

Tart, C. T., 'Geomagnetic effects on GESP: two studies', *Journal of the American Society for Psychical Research*, 1988; 82 (3): 193–215.

Tart, C. T., Initial Application of Mindfulness Extension Exercises in a Traditional Buddhist Meditation Retreat Setting, 1995 (unpublished: www.paradigm-sys.com/cttart).

Tedder, W. H. and Monty, M. L., 'Exploration of a long-distance PK: a conceptual replication of the influence on a biological system', in W. G. Roll et al. (eds.), *Research in Parapsychology 1980*, Metuchen, NJ: Scarecrow Press, 1981: 90–3.

Tiller, W., *Science and Human Transformation; Subtle Energies, Intentionality and Consciousness*, Walnut Creek, Calif.: Pavior Publications, 1997.

Tiller, W. et al., *Conscious Acts of Creation: The Emergency of a New Physics*, Walnut Creek, Calif.: Pavior Publishing, 2001.

Tiller, W. et al., *Some Science Adventures with Real Magic*, Walnut Creek, Calif.: Pavior Publishing, 2005.

Tiller, W. A., 'Subtle energies', *Science and Medicine*, 1999; 6 (3): 28–33.

Tiller, W. A. and Dibble, W. E. Jr., 'New experimental data revealing an unexpected dimension to materials science and engineering', *Material Research Innovation*, 2001; 5: 21–34.

Tiller, W. A. et al., 'Towards explaining anomalously large body voltage surges on exceptional subjects, Part I: The electrostatic approximation', *Journal of the Society for Scientific Exploration*, 1995; 9 (3): 331.

Tompkins, P. and Bird, C., *The Secret Life of Plants*, New York: Harper & Row, 1973.

Travis, F. and Wallace, R. K., 'Autonomic and EEG patterns during eyes-closed rest and Transcendental Meditation (TM) practice: the basis for a neural model of TM practice', *Consciousness and Cognition*, 1999; 8: 302–18.

Tromp, S. W., *Biometeorology*, London: Heyden, 1980.

Tschulakow, A. V. et al., 'A new approach to the memory of water', *Homeopathy*, 2005; 94: 241–7.

Tsushima, W. T., 'Treatment of phantom limb pain with EMG and temperature biofeedback: a case study', *American Journal of Clinical Biofeedback*, 1982; 5 (2): 150–3.

Ullman, M. et al., *Dream Telepathy: Experiments in ESP*, Jefferson, No.: McFarland, 1989.

Ursin, R. et al., 'Quantum teleportation across the Danube', *Nature*, 2004; 430: 849.

Utts, J., 'The significance of statistics in mind-matter research', *Journal of Scientific Exploration*, 1999; 13 (4): 615–38.

Vaiserman, A. M. et al., 'Human longevity: related to date of birth?' Abstract 9, Second International Symposium: Workshop on Chronoastrobiology and Chronotherapy, Tokyo Kasei University, Tokyo, Japan, November 2001.

Van Baalen, D. C. et al., 'Psychosocial correlates of "spontaneous" regression of cancer', *Humane Medicine*, April 1987.

Van Gyn, G. H. et al., 'Imagery as a method of enhancing transfer from training to performance', *Journal of Sport and Exercise Science*, 1990; 12: 366–75.

Van Wijk, E. P. A. and Van Wijk, R., 'The development of a bio-sensor for the state of consciousness in a human intentional healing ritual', *Journal of International Society of Life Information Science (ISLIS)*, 2002; 20 (2): 694–702.

Van Wijk, E. P. et al., 'Anatomic characterization of human ultra-weak photon emission in practitioners of Transcendental Meditation™ and control subjects', *Journal of Alternative and Complementary Medicine*, 2006; 12 (1): 31–8.

Van Wijk, R. and Van Wijk, E. P., 'The search for a biosensor as a witness of a human laying on of hands ritual', *Alternative Therapies in Health and Medicine*, 2003; 9 (2): 48–55.

Vedral, V., 'Entanglement hits the big time', *Nature*, 2003; 425: 28–9.

Vedral, V., 'Mean-field approximations and multipartite thermal correlations', *New Journal of Physics*, 2004; 6: 2–24.

Vedral, V., 'Quantifying entanglement', *Physical Review Letters*, 1997; 78 (12): 2275–9.

Vincent, J.-D., *The Biology of Emotions*, trans. J. Hughes. Cambridge, Mass.: Basil Blackwell, 1990.

Wackerman, J. et al., 'Correlations between brain electrical activities of two spatially separated human subjects', *Neuroscience Letters*, 2003; 336: 60–4.

Walker, E. H., *The Physics of Consciousness*, New York: Basic Books, 2000.

Wallace, B. A., 'The Buddhist tradition of Samatha: methods for refining and examining consciousness', *Journal of Consciousness Studies*, 1999; 6 (2–3): 175–88.

Walther, P., 'Quantum nonlocality obtained from local states by entanglement purification', *Physical Review Letters*, 2005; 94: 040504.

Watkins, G. K. and Watkins, A. M., 'Possible PK influence on the resuscitation of anesthetized mice', *Journal of Parapsychology*, 1971; 35: 257–72.

Watkins, G. K. et al., 'Further studies on the resuscitation of anesthetized mice', in W. G. Roll et al. (eds.), *Research in Parapsychology*, Metuchen, NJ: Scarecrow Press, 1973: 157–9.

Watt, C. et al., 'Exploring the limits of direct mental influence: two studies comparing "blocking" and "co-operating" strategies', *Journal for Scientific Exploration*, 1999; 13 (3): 515–35.

Weiman, C. E. and Cornell, E. A., 'Seventy years later: the creation of a Bose-Einstein condensate in an ultracold gas', *Lorentz Proceedings*, 1999; 52: 3–5.

Weinberg, R. S. et al., 'Effects of visuo-motor behavior rehearsal, relaxation, and imagery on karate performance', *Journal of Sport Psychology*, 1981; 3: 228–38.

Wells, R. and Klein, J., 'A replication of a "psychic healing" paradigm', *Journal of Parapsychology*, 1972; 36: 144–9.

West, M. A., 'Meditation and the EEG', *Psychological Medicine*, 1980; 10 (2): 369–75.

Wientjes, K. A., 'Mind-body techniques in wound healing', *Ostomy/Wound Management*, 2002; 48 (11): 62–7.

Wilkes, R. L. and Summers, J. J., 'Cognitions, mediating variables and strength performance', *Journal of Sport Psychology*, 1984; 6: 351–9.

Williams, B. J., 'Exploratory block analysis of field consciousness effects on global RNGs on September 11, 2001' (http://noosphere.princeton.edu/williams/GCP911.html).

Williams, P. and West, M., 'EEG responses to photic stimulation in persons experienced at meditation', *Electroencephalography and Clinical Neurophysiology*, 1975; 39 (5): 519–22.

Winton, J., 'New functions for electrical signals in plants', *New Phytologist*, 2004; 161: 607–10.

Wolf, F. A., *Mind into Matter: A New Alchemy of Science and Spirit*, Needham, Mass.: Moment Point Press, 2000.

Yue, G. H. and Cole, K. J., 'Strength increases from the motor program; comparison of training with maximal voluntary and imagined muscle contractions', *Journal of Neurophysiology*, 1992; 67: 114–23.

Zeilinger, A., 'Probing the limits of the quantum world', *Physics World*, March 2005 (online journal: http://www.physicsweb.org/articles/world/18/3/5/1).

Zeilinger, A., 'Quantum teleportation', *Scientific American*, April 2000: 32–41.

Zimmerman, J., 'New technologies detect effects in healing hands', *Brain/Mind Bulletin*, 1985; 10 (2): 20–3.

Zohar, D., *The Quantum Self*, London: Bloomsbury Press, 1991.

Zwierlein, M. W. et al., 'Observation of Bose-Einstein condensation of molecules', *Physical Review Letters*, 2003; 91: 250401.

Useful websites

www.biomindsuperpowers.com: Ingo Swann's Superpowers of the Human Bio-mind
www.fourmilab.ch/rpkp/bierman-metaanalysis.html
www.laurentian.ca/Neursci/_people/Persinger.htm
www.lifescientists.de: official website of the IIB.
www.officeofprayerresearch.org
www.spiritualbrain.com
www.wholistichealingresearch.com

Index